Security & Intelligence

Also from Westphalia Press
westphaliapress.org

S&I Security & Intelligence

Volume 10, Number 1 • Spring 2025

Carter Matherly , Matthew Loux & Jim Burch, editors

Westphalia Press
An imprint of Policy Studies Organization

SECURITY & INTELLIGENCE
VOLUME 10, NUMBER 1 • SPRING 2025

All Rights Reserved © 2025 by Policy Studies Organization

Westphalia Press
An imprint of Policy Studies Organization
1367 Connecticut Avenue NW
Washington, D.C. 20036
info@ipsonet.org

ISBN: 978-1-63723-677-2

Interior design by Jeffrey Barnes
jbarnesbook.design

Daniel Gutierrez-Sandoval, Executive Director
PSO and Westphalia Press

Updated material and comments on this edition
can be found at the Westphalia Press website:
www.westphaliapress.org

Security and Intelligence strives to be the source for research on global security and intelligence matters. As the global threat-scape evolves over time, *SI* is evolving to keep pace. The journal is enhancing its academic edge, impact, and reach. We are working to build stronger bridges between senior leaders, academics, and practitioners. In addition to new content that advances the global discussion of security and intelligence, readers can anticipate more special issues with a focus on current security concerns.

Security and Intelligence is one of five journals sponsored by American Public University and published by Policy Studies Organization. The *International Journal of Online Educational Resources* (IJOER) publishes academic research with an emphasis on representing Open Educational Resources in teaching, learning, scholarship, and policy. The *Journal of Online Learning Research and Practice* (JOLRAP) publishes articles that focus on aspects related to virtual instruction, technology integration, data, ethics, privacy, leadership, and more. *Space Education and Strategic Applications* (SESA) journal encourages the publication of advances in space research, education, and applications. And lastly, *The Saber and Scroll* is a student- and alumni-led journal that publishes a variety of research on history and military history topics, book reviews, and exhibit/museum reviews. Please visit https://www.apus.edu/academic-community/journals/index for more information on each journal.

> *A very special thank you to American Public University and the Policy Studies Organization for your generous and continued support of SI.*

If you have research, notes, concepts, or ideas that you want to share, please do not hesitate to reach out with your submission! Our editorial team is always available to support new authors seeking to make an impact on the industry. Please visit us on Scholastica, https://gsis.scholasticahq.com/for-authors, for specifics on submissions.

Security and Intelligence

Security and Intelligence is published by The Policy Studies Organization on behalf of American Public University System. *SI* is licensed under a Creative Commons Attribution - NonCommercial - NoDerivatives 4.0 International License.

Aims and Scope. SI is a bi-annual, peer-reviewed, open-access publication designed to provide a forum for the academic community and the community of practitioners to engage in dialogue about contemporary global security and intelligence issues. The journal welcomes contributions on a broad range of intelligence and security issues, and from across the methodological and theoretical spectrum.

The journal especially encourages submissions that recognize the multidisciplinary nature of intelligence and security studies, and that draw on insights from a variety of fields to advance our understanding of important current intelligence and security issues. In keeping with the desire to help bridge the gap between academics and practitioners, the journal also invites articles about current intelligence and security-related matters from a practitioner perspective. In particular, *SI* is interested in publishing informed perspectives on current intelligence and security-related matters.

SI welcomes the submission of original empirical research, critical analysis, policy analysis, research notes, and book reviews. Papers and research notes that explicitly demonstrate how a multidisciplinary approach enhances theoretical and practical understanding of intelligence and security matters are especially welcome. Please visit:

https://www. ipsonet.org/publications/open-access/gsis/instructions-for-authors

Security & Intelligence
Volume 10, Number 1 • Spring 2025
© 2025 Policy Studies Organization

(cont'd.)

Book Review

Editorial Welcome to the Tenth Volume

G reetings! Welcome to the first issue of our milestone 10th volume. In this issue of *Security and Intelligence* (S&I), scholars examine a broad array of topics from justified belief to modern warfare and peacekeeping risks.

Our tenth volume opens with a Graduate Lectern worthy of attention. Graduate student Sherry Dingman, with her professor Dr. Kathryn Lambert, brings the soap opera story of La Familia Michoacan (LFM), a Mexican cartel, to academic light. The research highlights the adaptability of criminal organizations to ensure survival.

Our first research article of the issue is "Justification (Justified True Belief) in the Intelligence Context: Testimony and Justification Defeaters" from Linda Johansson, in which they examine "justified belief" while considering justified defeaters in argument maps/trees to assess testimony and evaluate sources in the intelligence community. We then turn our attention to a modern warfare discussion in "Kinetics to Keyboards: The Future of the Special Operations Force Operator Post Global War on Terror" by AJ Rutherford focusing on cybersecurity, digital forensics, and network penetration training, along with collaboration with government, private, and academic entities, providing policy recommendations for enhanced training. In the article "Silent Sacrifices: Political Predictors of UN Peacekeeper Fatalities," Kathryn M. Lambert then exposes the seldom discussed risks to peacekeeping missions. International cooperation is explored in "The Future of FVEY: Could France Replace New Zealand as an Intelligence-Sharing Partner?" by Victoria J. Sengelman, which delves into the importance of international secrets-sharing practices and partnerships in the fight against terrorism and global security. Sengelman examines a foreign intelligence-sharing partnership with New Zealand, particularly focusing on the potential formation of a partnership with France. "The Cross-Border Cooperation Between Colombia, Peru, and Venezuela" by Laura J. Gender & David J. Kritz explores the effects of geopolitics, political conflicts, and resulting transnational criminal activity on cross-border cooperation and intraregional communication among Colombia, Peru, and Venezuela. "Natural Language Processing for the Intelligence Analysis Process" by Brandon Morad investigates the potential benefits and challenges of integrating Natural Language Processing (NLP) into the US Intelligence Community's (IC)

doi: 10.18278/si.10.1.1

intelligence analysis process. Morad's qualitative analysis aims to explore the introduction of NLP into the intelligence analysis process. Finally, "Active Shooter Awareness and Preparedness in Soft Target Scenarios" by Joshua E. Lane examines preparedness and awareness as crucial factors in educating people about the impact of active shooters, which Lane identifies as increasingly significant, resulting in a greater need for awareness and preparedness. Our volume closes out with a detailed and authoritative review of *Terrorist Minds: The Psychology of Violent Extremism from Al-Qaeda to the Far Right*.

Thank you for joining us on our journey crossing combat lines, examining the intelligence lines of nations in the First World War, exploring social influences on food insecurities, and learning from past analytical mistakes with lessons learned and new analytical approaches. It is a privilege to bring another issue of Security and Intelligence to our readers and the community.

Carter Matherly, Ph.D.
Co-Editor in Chief

Matthew Loux, D.M.
Co-Editor in Chief

Jim Burch, D.M.
Associate Editor

Editorial de bienvenida al décimo volumen

¡Saludos! Le damos la bienvenida al primer número de nuestro décimo volumen, un hito histórico. En este número de Security and Intelligence (S&I), investigadores examinan una amplia gama de temas, desde la creencia justificada hasta los riesgos de la guerra moderna y el mantenimiento de la paz.

Nuestro décimo volumen abre con un atril de posgrado digno de atención. La estudiante de posgrado Sherry Dingman, junto con su profesora, la Dra. Kathryn Lambert, saca a la luz académica la historia de La Familia Michoacán (LFM), un cártel mexicano. La investigación destaca la adaptabilidad de las organizaciones criminales para asegurar su supervivencia.

Nuestro primer artículo de investigación de este número es "Justificación (Creencia Verdadera Justificada) en el Contexto de Inteligencia - Testimonio y Derrotadores de la Justificación" de Linda Johansson. En él, se examina la "creencia justificada" considerando los derrotadores justificados en mapas/árboles de argumentos para evaluar el testimonio y las fuentes de inteligencia. Posteriormente, nos centramos en un debate sobre la guerra moderna en "De la cinética a los teclados: El futuro del operador de la Fuerza de Operaciones Especiales tras la Guerra Global contra el Terror", de AJ Rutherford, que se centra en la ciberseguridad, la informática forense y la formación en penetración de redes, junto con la colaboración con entidades gubernamentales, privadas y académicas, ofreciendo recomendaciones de políticas para una formación mejorada. En el artículo "Sacrificios Silenciosos: Predictores Políticos de las Muertes de Personal de Paz de la ONU", Kathryn M. Lambert expone los riesgos poco abordados para las misiones de mantenimiento de la paz. La cooperación internacional se explora en "El Futuro de FVEY: ¿Podría Francia Reemplazar a Nueva Zelanda como Socio para el Intercambio de Inteligencia?", por Victoria J. Sengelman, que profundiza en la importancia de las prácticas y asociaciones internacionales de intercambio de secretos en la lucha contra el terrorismo y la seguridad global. Sengelman examina una asociación de intercambio de inteligencia extranjera con Nueva Zelanda, centrándose particularmente en la posible formación de una asociación con Francia. "La cooperación transfronteriza entre Colombia, Perú y Venezuela" por Laura J. Gender y David J. Kritz explora los efectos de la geopolítica, los conflictos políticos y la actividad criminal transnacional resultante en la cooperación transfronteriza y la comunicación intrarregional entre Colombia, Perú y Venezuela. "Procesamiento del lenguaje natural para el proceso de análisis de inteligencia" por Brandon Morad investiga los posibles beneficios y desafíos de integrar el procesamiento del lenguaje natural (PLN) en el proceso de análisis de inteligencia de la comunidad de inteligencia de los Estados Unidos (CI). El análisis cualitativo de Morad tiene como objetivo explorar la introducción del PLN en el proceso de análisis de inteligencia. Finalmente, "Active Shooter Awareness and Preparedness in Soft Target Scenarios" de Joshua E. Lane examina la preparación y la concienciación como factores cruciales para educar a la gente sobre el impacto de los tiradores activos, que Lane identifica como cada vez más significativo, lo que resulta en una mayor necesidad de concienciación y preparación. Nuestro volumen concluye con una reseña detallada y fidedigna de Mentes terroristas: La psicología del extremismo violento desde Al Qaeda hasta la extrema derecha.

Gracias por acompañarnos en nuestro viaje, cruzando las líneas de combate, examinando las líneas de inteligencia de las naciones durante la Primera Guerra Mundial, explorando las influencias sociales en la inseguridad alimentaria y aprendiendo de errores analíticos pasados con lecciones aprendidas y nuevos enfoques analíticos. Es un privilegio presentar un nuevo número de Seguridad e Inteligencia a nuestros lectores y a la comunidad.

编者按：《安全与情报》杂志第十卷

各位读者，您好！欢迎阅读《安全与情报》(S&I)杂志具有里程碑意义的第十卷第一期。本期中，学者们分析了从确证信念到现代战争和维和风险等一系列广泛主题。

第十卷的第一篇文章来自"研究生讲台"版块。研究生Sherry Dingman和她的教授Kathryn Lambert博士将墨西哥卡特尔组织"米却肯家族"(LFM)的肥皂剧故事带入学术界。文章强调了犯罪组织为确保生存而采取的适应性。

本期收录的第一篇研究文章是Linda Johansson撰写的《情报语境中的证成（确证的真实信念）——证词与证成破坏因素》(Justification (Justified True Belief) in the Intelligence Context - Testimony and Justification Defeaters)。文章探讨了"确证信念"，并运用论证图/树中的确证破坏因素来评估证词并评价情报界的消息来源。之后，我们将注意力转向AJ Rutherford撰写的《从动力学到键盘：全球反恐战争(GWOT)后特种作战部队(SOF)操作者的未来》(Kinetics to Keyboards: The Future of the Special Operations Force Operator Post Global War on Terror)一文中关于现代战争的讨论，该文聚焦于网络安全、数字取证和网络渗透训练，以及与政府、私营和学术实体的合作，为加强训练提供了政策建议。Kathryn M. Lambert在《无声的牺牲：联合国维和人员伤亡的政治预测因素》(Silent Sacrifices: Political Predictors of UN Peacekeeper Fatalities)一文中揭示了维和特派团鲜为人知的风险。Victoria J. Sengelman撰写的《五眼联盟的未来：法国能否取代新西兰成为情报共享伙伴？》(The Future of FVEY: Could France Replace New Zealand as an Intelligence-Sharing Partner?)探讨了国际合作，文章研究了国际机密共享实践和伙伴关系在打击恐怖主义和维护全球安全方面的重要性。Sengelman分析了与新西兰的外国情报共享伙伴关系，聚焦于美法两国建立伙伴关系的可能性。《哥伦比亚、秘鲁和委内瑞拉之间的跨境合作》(The Cross-Border Cooperation Between Colombia, Peru, and Venezuela)由Laura J. Gender和David J. Kritz撰写，探讨了地缘政治、政治冲突及其导致的跨国犯罪活动对哥伦比亚、秘鲁和委内瑞拉之间跨境合作和区域内交流的影响。Brandon Morad撰写的《情报分析过程中的自然语言处理》(Natural Language Processing for the Intelligence Analysis Process)探讨了"将自然语言处理(NLP)整合到美国情报界(IC)情报分析过程"一事的潜在优势和挑战。Morad的定性分析旨在探究"将NLP引入情报分析过程"。最后，Joshua E. Lane撰写的《软目标场景中的现场行凶枪手意识和准备》(Active Shooter Awareness and Preparation in Soft Target Scenarios)分析了准备和意识这两个关键因素，以教育人们了解现场行凶枪手的影响，Lane认为这一点日益重要，它导致对意识和准备的需求也日益增加。本卷的最后一篇文章是一篇详细而权威的书评，书评对象是《恐怖分子的心智：从基地组织到极右翼的暴力极端主义心理学》(Terrorist Minds: The Psychology of Violent Extremism from Al-Qaeda to the Far Right)。

感谢您与我们一起跨越战线，分析第一次世界大战期间各国的情报线，探究社会对粮食不安全的影响，并从过去的分析失误中汲取经验教训，并探索新的分析方法。我们很荣幸能为读者和情报界带来新一期的《安全与情报》杂志。

The Evolution of La Familia Michoacan: A Case Study

Sherry Dingman

School of Security and Global Studies, American Public University System

Dr. Kathryn Lambert

Abstract

The story of La Familia Michoacan (LFM), like a Mexican soap opera, features romantical betrayals, greed, revenge and violence. It is a story of an organization fragmenting, forming alliances, and adapting to its environment. LFM proclaimed an ideology rooted in fundamental evangelicalism, claiming to protect the citizens of Michoacan from outsiders and corrupt politicians. The organization exploited the weakness of political institutions, spoke the language of an insurgency, passed out humanitarian aid and even enforced mask mandates during COVID-19. LFM exercised control over the largest port in Mexico, Lazaro Cardenas, from whence it obtained cocaine from South American and chemicals for synthesizing drugs. In its newest iteration, La Nueva Familia Michoacana (NFM), it is engaged in a deadly struggle with the Jalisco New Generation Cartel (CJNG) for control of territory, port Lazaro Cardenas and the lucrative avocado industry. Both groups have been declared Foreign Terrorist Organizations by the U.S. State Department. A major lesson to be learned from LFM is the adaptability of criminal organizations. Designating cartels as terrorist organizations may reduce corruption in the Mexican government, but the criminals are likely to continue supplying consumer demand. How else can the sons of impoverished Mexican peasants hope to rule over global enterprises worth billions?

Keywords: Case Study; La Familia Michoacan; Mexican Cartels; Organised Crime; Terrorist Organizations

La evolución de La Familia Michoacán: un estudio de caso

Resumen

La historia de La Familia Michoacana (LFM), como una telenovela mexicana, presenta traiciones románticas, avaricia, venganza y

 doi: 10.18278/si.10.1.2

violencia. Es la historia de una organización que se fragmenta, forma alianzas y se adapta a su entorno. LFM proclamó una ideología arraigada en el evangelicalismo fundamental, afirmando proteger a los ciudadanos de Michoacán de forasteros y políticos corruptos. La organización explotó la debilidad de las instituciones políticas, habló el lenguaje de una insurgencia, distribuyó ayuda humanitaria e incluso impuso el uso obligatorio de cubrebocas durante la COVID-19. LFM controlaba el puerto más grande de México, Lázaro Cárdenas, de donde obtenía cocaína sudamericana y químicos para sintetizar drogas. En su versión más reciente, La Nueva Familia Michoacana (NFM), libra una lucha a muerte con el Cártel Jalisco Nueva Generación (CJNG) por el control del territorio, el puerto de Lázaro Cárdenas y la lucrativa industria del aguacate. Ambos grupos han sido declarados Organizaciones Terroristas Extranjeras por el Departamento de Estado de Estados Unidos. Una lección importante que se puede aprender de LFM es la adaptabilidad de las organizaciones criminales. Designar a los cárteles como organizaciones terroristas puede reducir la corrupción en el gobierno mexicano, pero es probable que los criminales sigan abasteciendo la demanda de los consumidores. ¿De qué otra manera podrían los hijos de campesinos mexicanos empobrecidos aspirar a gobernar empresas globales que valen miles de millones?

Palabras clave: Estudio de caso; La Familia Michoacán; Cárteles mexicanos; Crimen organizado; Organizaciones terroristas

米却肯家族的演变：一项案例研究

摘要

米却肯家族(LFM)的故事如同一部墨西哥肥皂剧，充斥着爱情背叛、贪婪、复仇和暴力。这是一个关于组织分裂、结盟并适应环境的故事。LFM宣扬一种根植于原教旨福音派的意识形态，声称要保护米却肯州公民免受外来者和腐败政客的侵害。该组织利用政治制度的弱点，使用叛乱的语言，发放人道主义援助，甚至在新冠疫情期间实行口罩强制令。LFM控制着墨西哥最大的港口拉萨罗卡德纳斯，并从那里获取来自南美的可卡因和用于合成毒品的化学品。LFM最近已发展为新米却肯家族(NFM)，该组织正与哈利斯科新一代卡特尔(CJNG)展开殊死搏斗，争夺领土、拉萨罗卡德纳斯港口以及利润丰厚的鳄梨产业的控制权。这两个组织均已被美国国务院列为外国恐怖组织。从LFM事件中汲取的一个重要教训是犯罪组织的适应性。将卡特尔指定为恐怖组织，或许能减少墨西哥政府的腐败，但犯罪分子很可能会继续满足消费者的需求。否则，这些贫困墨西哥农民的子弟又怎能指望统治价值数十亿美元的全球企业呢？

关键词：案例研究，米却肯家族，墨西哥卡特尔，有组织犯罪，恐怖组织

An Analysis of La Familia Michoacán (LFM*)*

This paper considers the origin and development of La Familia Michoacan (LFM) as it adapted and survived. Fragmenting and forming alliances, the organization has outlived leaders who were imprisoned or executed. LFM has all the elements of a Mexican soap opera—romantic entanglements, plot twists, betrayals, and memorable characters. Early in its history, LFM embraced a religious ideology that justified brutality as divine justice. LFM is the story of poor young men in rural Mexico participating in the global economy by growing marijuana, poppies, cooking methamphetamine and cashing in on a rule allowing Michoacan avocados to be exported to the U.S.[1] In the 1970s, avocado orchards served as a means for laundering drug money. Now the crop is so profitable that cartels battle to control its production and distribution (Shortell, 2024; Global Initiative Against Organized Crime, 2024). A recent editorial in *The New York Times* cautioned against designating the Mexican cartels as foreign ter-

rorist organizations as this could force Americans to stop doing business in Mexico. The piece cited avocado farming as an example of how criminals are "embedded in the legal economy" (Abi-Habib and Romero, 2025). It might be naive to call the industry "legal" once cartels get involved. Not content with extorting growers, the criminals are expanding production by killing people and displacing others from protected lands. Over 70,000 acres have been illegally deforested to establish avocado orchards. Starting a new orchard requires getting a permit that often involves "persuading" an official. About 80 percent of avocado orchards in Michoacan were established illegally (Appleby, 2024). Jalisco New Generation Cartel (CJNG), LFM and others are battling to control the industry. The price of opium dropped drastically after the debut of synthetic opioids, so avocados are replacing poppies as a revenue stream for cartels (Rainsford, 2019).

Methods

Studying transnational criminal organizations (TCOs) presents serious methodological challeng-

---

[1] https://www.federalregister.gov/documents/2004/11/30/04-26336/mexican-avocado-import-program

es because they are secretive by nature and law enforcement is political by necessity. Hard data on drug trafficking is scarce, TCOs publish no quarterly reports on profitability. Any public communications by TCOs are likely to be motivated by propaganda objectives (Kenney, 2007). Law enforcement agencies must be concerned about annual budgets and presumably are inclined to report on TCOs accordingly. Many scholarly articles and news reports about LFM are published in Spanish, making them accessible only with translation (thank you Google). Fortunately, academics from Michoacan, like Dr. Salvador Maldonado Aranda, have published in English (2013). LFM has gained sufficient notoriety to be the subject of a thesis (De Amicis, 2010), a monograph (Grayson, 2011), a chapter in the *Handbook of Homeland Security* (Corcoran, 2023), scholarly articles (for example by Bunker and Sullivan, 2020; Atuesta and Pérez-Dávila, 2018; Flanigan, 2014; Aguirre & Herrera, 2013; Logan and Sullivan, 2009), mention in books by professors (Smith, 2025) and journalists (Hernandez, 2013; Grillo, 2011), reported about by think tanks (Council on Foreign Relations, InSight Crime, Organized Crime and Corruption Project, Wilson Center), various agencies of the U.S. government, and by the media (Flannery, 2023; Ferri, 2022; Fainaru and Booth, 2009). From all these sources and others, this case study of LFM was crafted.

Results

History

The origins of La Familia Michoacán (LFM) are variously attributed to a vigilante group in the 1980s or a cooperative venture between marijuana and poppy growers. Grayson (2010) describes how the group began with alliances and factions of El Milenio, Los Zetas and La Empresa. FLM has fragmented and formed alliances since its inception, evolving and adapting to survive (Atuesta and Pérez-Dávila, 2018).

The Milenio Cartel began with a family of avocado growers in the state of Michoacan, the Velencia family, who diversified into marijuana and poppies by the 1950s. By the 1990s, the family had connections with Pablo Escobar to distribute cocaine. With the dawn of a new century at hand, the family business was named the Milenio Cartel. Chinese Mexican businessman Zhenli Ye Gon oversaw the importation of tons of pseudoephedrine and ephedrine into Mexico for Unimed Pharm Chem Mexico. He helped the Velencia family by diverting precursor chemicals so they could produce methamphetamine (BBC, 2016).

Carlos Rosales Mendoza who worked for the Velencia family was tasked with establishing a business collaboration with the Gulf Cartel to move Milenio drugs to the U.S. border. At this time the leader of the Gulf Cartel, Osiel Cárdenas Guillén, was perhaps the most powerful drug lord in Mexico. Cárdenas recruited former Mexican

Special Forces to serve as his security, the Zetas. Rosales and his men were trained by Gulf Cartel's Zetas. Interpersonal ties fostered the business relationship between the Milenio and Gulf Cartels. The Valencia brothers, Cárdenas, and Rosales were friends. One of the men working for the Milenio Cartels was Nemesio Rubén Oseguera Cervantes (El Mencho). El Mencho, raised in a poor family, dropped out of school in the fifth grade to cultivate avocadoes. He made repeated trip to the U.S. from whence he was deported before joining the police force in Jalisco. By the time he met Rosales, El Mencho was a member of the assassination squad for the Milenio Cartel. He was also married to one of the sisters of its leader, Armando Valencia. El Mencho introduced Rosales to his cousin Ines Oseguera. Rosales fell in love with her and the couple had a child together (Infobae, 2019).

The trouble that led to the formation of LFM began when Armando Valencia, leader of the Milenio Cartel, and brother-in-law of El Mencho, had an affair with Ines Oseguera (Infobae, 2019). Rosales was furious about the betrayal and told a reporter he became "a love-struck demon" (Infobae, 2019). At the time of the affair, Rosales was married to another woman (Grayson, 2010). In February 2002, Rosales executed a revenge attack, killing Armando Valencia's nephew and three others, but not Armando himself. Osiel Cárdenas of the Gulf Cartel, friend of Rosales, detected an opportunity in the rift. To support Rosales in his quest for revenge, Cárdenas dispatched Zeta hitmen in 2002 to eliminate the Valencia family

(Grayson, 2010). The Valencia family sought help from "El Chapo" and his Sinaloa Cartel. Those loyal to Rosales joined him in forming La Empresa (the business) a subsidiary of the Gulf Cartel (Greyson, 2010.) Armando Valencia was captured in 2003 and sentenced to 47 years in prison. The Milenio Cartel fractured, part merged with the Sinaloa Cartel and another, part reincarnated as Jalisco New Generation Cartel under the leadership of El Mencho (CJNG) (*InSight Crime*, 2024).

Ever loyal to Cardenas, Rosales attacked the Apatzingán prison in 2003 to free members of the Gulf Cartel. In an operation that lasted less than seven minutes, 60 armed men dressed in military and police uniforms entered the prison. Authorities could not determine whether the men were members of Los Zetas or merely trained by them. At the time, Rosales was considered the leader of the Gulf Cartel in Michoacan state (*Washington Times*, 2004).

Another woman allegedly in a relationship with Rosales was La Jefa, a drug dealer and sex trafficker. Her body was discovered decapitated and sliced open from her chest to stomach (she had been pregnant). Investigators said she had been dealing drugs outside her assigned area. The month prior to this execution, 88 people had been murdered in the state of Michoacan, four decapitated, five dismembered, and 79 shot (Arrieta 2019). Rosales sought vengeance on those he blamed for her death (Arrieta 2019). The public debut of LFM came three days later when five severed heads were tossed onto a dance floor with a message that LFM did not kill women (Grayson, 2010.) The heads belong to Zetas whom LFM found guilty of raping and murdering a pregnant prostitute for refusing to have sex with them (Smith ,2021). She was apparently in a relationship with a member of the newly formed LFM.

After Cardenas Guillen was arrested in 2003, the Zetas moved towards independence from the Gulf Cartel. They wanted control of Michoacan and its port. Zetas specialized in violence, using military tactics to control territory by force. Their primary business was extortion (Smith, 2021). Osiel Cardenas Guillen continued to run the Gulf Cartel from inside a Mexican maximum-security prison. In 2007, Cardenas was extradited to the U.S. and sentenced to 25 years. He was turned over to Mexican authorities in 2024.[2] Rosales was arrested in October of 2004 after an unsuccessful attempt to free his friend from prison.

Rosales served ten years in a Mexican prison for attempting to free Cardenas and other crimes. During his imprisonment other members took over running LFM included Nazario Moreno, Jesus Mendez, and Servando Gomez (Infobae, 2019). Rosales was released from prison in May of 2014. About a year later his release, his body was found with four others in a parking lot. They had been tortured before their execution. According to an interview with a state prosecutor, Rosales

had attended a meeting of self-defense leaders. At that meeting he had a disagreement with Ignacio Andrade about starting a new cartel in Michoacan. Andrade allegedly ordered the abduction and execution of Rosales (Davis, 2015). In 2009, Ignacio Renteria Andrade was named in an indictment as a member of La Familia which had been designated a significant foreign narcotics trafficking organization under the Foreign Narcotics Kingpin Designation Act.[3]

During Rosales' years in prison, LFM fragmented into competing groups. Nazario Moreno González took leadership and declared independence from the Gulf Cartel. He sought to rid Michoacana of Los Zetas and other external influences. It was Moreno who provided LFM with a religious ideology that espoused a duty to protect Michoacan. Moreno, a fan of Eldredge's book *Wild at Heart*, embraced his own brand of evangelical fundamentalism. Moreno's book *Pensamientos* (Thoughts) was required reading for LFM initiates. The cartel paid to print and distribute it to rural schools in the state. The book offered inspirational religious sayings and revolutionary slogans. It resonated with Moreno's audience and gave LFM members a higher purpose. Monroe offered a rationale for brutality in the service of divine justice. He gave LFM its unique brand. When he was reported to have died in a shoot-out in 2010, he gained the reputation as a saint because people claimed to have seen him alive. The

---

2   https://www.ice.gov/news/releases/former-leader-gulf-cartel-removed-after-being-released-prison

3   https://www.justice.gov/archive/usao/nys/pressreleases/October09/lafamiliaindictmentpr.pdf

2010 report of his death was premature. Monroe continued in leadership until 2014 when his death in another shoot out was confirmed with fingerprints (Grayson, 2011).

A faction of LFM calling itself the Knights Templar (KT) was founded in 2011 after Monroe's first reputed death. On public banners, KT declared it would be taking over altruistic activities previously done by LFM. After Moreno was confirmed dead in 2014, more members of LFM aligned themselves with the lieutenants who had formed the KT. The group maintained a version of the ideology espoused by Moreno, holding initiation rites using white robes and red crosses (Lomnitz, 2019). One of the lieutenants who lead the Templars, Servando Gomez, was arrested in 2015. The last of the founding members of KC, Ignacio Renteria Andrade, was arrested by Mexican security forces in Michoacán in 2017.

Jose de Jesus Mendez Vargas remained in control of a smaller faction of the original LFM who had not joined KT. Mendez was arrested in 2011. He claimed LFM would not survive after his arrest, but he was wrong (Hopkins, 2011). Mendez was extradited to the U.S. in February 2025. One faction of LFM has had links to the apocalyptic New Jerusalem movement founded by a defrocked Catholic priest (Grayson, 2010). Eventually factions of LFM, remnants of KTC, the Sinaloa Cartel with its Milenio subsidiary, and Gulf Cartel formed a joint fighting force called the United Cartels or La Resistencia. The force was dedicated to fighting Los Zetas.

When its leaders died, the Milenio Cartel fragmented, and one part became CJNG under the leadership of El Mencho. CJNG portrayed itself as protectors of the people, perhaps infringing on the LFM brand (Jones, 2018). Leaders of the KT were on their way to a meeting to discuss an alliance with CJNG when they were assassinated (Atuesta and Pérez-Dávila, 2018). The group that carried out the attack was Nueva Familia Michoacana (NFM) now headed by the Olascoaga brothers, Johnny Hurtado Olascoaga and Jose Alfredo Hurtado. With the capture of Mendez, the brothers had taken over leadership of LFM in 2014 (Mayen, 2023). The U.S. government is offering an $8 million dollar reward for the brothers.

The "Nueva" Familia Michoacana (NFM) announced its existence on banners hung around Michoacana in February 2016. NFM declared it would drive CJNG and other extortionists, kidnappers, robbers, rapists and assassins out of the region. NFM took control of an area called Tierra Caliente, straddling Michoacan, Guerrero and Mexico states. NFM formed an alliance with a remnant of the KT, Los Viagras led by Moreno's nephew. Los Viagras began as a self-defense group, but when the Mexican government demanded it disarm in 2014, it resisted and expanded into drug trafficking. KT was mostly defeated by vigilante militias (Arrieta, 2019; Bargent, 2016). Self-defense groups sometimes resorted to drug trafficking to obtain weapons.

In 2019, CJNG tortured and killed 19 members of Los Viagra, displaying their dismembered bodies in the city of Uruapan, in the center of the avocado export region. The dead included three women and CJNG left a message: "Lovely people, carry on with your routines. Be patriotic and kill a Viagra" (Henkin 2024, 1). Thirteen police officers died in an ambush on a convoy in an attack declared "an affront to the Mexican State" likely carried out by LFM/NFM in March 2021 (Agren, 2021). The attack took place in Mexico state, which surrounds Mexico City on three sides. Authorities have identified 26 criminal groups active in the state of Mexico (Agren, 2021). In December 2022, a firefight killed eleven alleged members of LFM. A non-aggression pact with CJNG collapsed in March of 2022 when 20 people connected with CJNG were killed in an attack. In June of 2022, Eduardo Hernández Vera, who led a remnant of LFM was executed by order of the Olascoaga brothers over a disagreement within the organization (*The Universal*, 2022).

Ten cartel members and four farmers were killed in December of 2023 in a battle that broke out after a meeting called to discuss fees LFM was assessing the farmers (Guillen, 2023). FFM is battling CJNG for control of territory in Michoacan and nearby states (Appleby, 2023).

For a time, LFM had its own bible and operated under a code of ethics, purporting to exist for the good of the people. Over the decades FFM fractured, spawned new groups, and endured deaths and incarcerations of leaders. Its newest iteration, NFM is allegedly involved in trafficking fentanyl, cocaine, methamphetamine and migrant smuggling (U.S. Dept of State 2024; U.S. Dept of Treasury, 2022). In February 2025, it was designated a foreign terrorist organization.

Geography

Michoacan is prime real estate for agriculture and home to the major Pacific port of Lazaro Cardenas. The port's location facilitates the transit of drugs, including cocaine from South America, synthetic opioids from China, and chemicals from Asia needed to make synthetic drugs. Michoacan state has a long history of growing poppies and marijuana (Aranda, 2013). The region was left untouched by federal development for decades except for construction of small airstrips to serve remote communities. Public services and representatives of the federal government were seldom seen in Michoacan. Eventually, material goods from drug trafficking money appeared, satellite televisions, and luxury vehicles. The federal military came to visit to destroy the poppy and marijuana crops that the people grew to survive economically. Southern Michoacán was once synonymous with the frontier because it was nearly inaccessible (Grayson, 2015). The federal government finally began incorporating the region into the Mexican economy by investing in infrastructure. The same roads that made it possible to export crops facilitated drug trafficking, including the coastal highway built in the 1980s to serve the port of Lazaro Cardenas.

Culturally, the frontier nature of the region gave rise to a kind of rugged individualism, strong family ties, and opposition to the central government (Aranda, 2013). The region cultivates avocados, an industry that served as a means of laundering money in the days of the Valencia family (Aranda 2013). Seventy-nine percent of the people of Michoacán lived in poverty and men between 18 and 45 went to the U.S. to find work. Emigrating became far more difficult in 1986 due to U.S. legislation. The law did not undo families ties to kin already in the U.S. who could help distribute drugs. The North American Free Trade Agreement (NAFTA) dealt a blow to the region as did the economic recession of 2008–2010 which left thousands of young men unemployed (Grayson, 2010). Because of cultural norms, women were excluded from drug trafficking careers (this is changing now), but men gravitated toward it and its allied industries (Aranda, 2013). LFM offered the promise of job security to any man willing to commit to two months of extensive bible study and hearing evangelistic messages. Members were recruited out of rehabilitation programs. The men were given well paying positions to traffic drugs to the U.S. (Grayson, 2010). Methamphetamine was produced exclusively for export to the U.S. and any distribution locally was prohibited by LFM (Grayson, 2010). The crack down on meth labs and chemical precursors in the U.S. created a business opportunity for LFM (Wilson Center, 2000). Members who joined the cartel were doing the Lord's work, safeguarding women and chil-

dren, combating external threats, and serving the people of Michoacan.

When the Mexican government decentralized, it granted increased authority to states and municipalities. The state government made the police, rather than the army responsible for public safety. Military intervention to burn crops in Michoacan had created indignation on the part of many residents of the state (Aranda, 2013).

The Valencia family formed an alliance with the Amezcua brothers to control the Pacific coast of Mexico. The Amezcua family was in the business of producing methamphetamine and the Valencia family joined them, investing profits into avocado orchards and real estate. The Sinaloa and Gulf Cartels were fighting for control over the Nuevo Laredo region. In this dispute, the Valencia family sided with El Chapo and his Sinaloa Cartel. Rosales, a former associate of the Valencia family, sided with the Gulf Cartel. The exceptional clash between the cartels was for control of major transportation routes and drug production in Michoacán. The Valencia deployed men trained by Guatemalan ex-special forces who had served in the Mexican military. The Gulf Cartel had the Zetas, former Mexican special forces. Battles were brutally violent (Aranda, 2013).

LFM was responsible for killing hundreds of people during its first three years of existence (Grayson, 2010). LFM was organized in para-military style, deployed guerrilla units, and was fueled by an ideology of social justice. The combat arm of LFM trained with

15

the Zetas before Armando Valencia betrayed Rosales by sleeping with his woman.

The President of Mexico from 2006–2012, Felipe Calderon, was from Michoacan state. He started a crusade to combat drug trafficking. He sent 7,000 military and police to patrol Michoacan. While the campaign was successful in catching some of the leaders of criminal organizations, government forces were accused of torture and other human rights abuses. The federal government claimed LFM had ties to the state government. LFM countered with the accusations that the Federal Police protected cartels (Smith, 2021). Genaro García Luna, the Mexican Secretary of Security under Calderon was paid millions to assist the Sinaloa Cartel (Hackbarth, 2020). Garcia was eventually convicted in New York in 2023 (CBS, 2024).[4]

In May of 2009, 29 officials in Michoacan were accused of corruption and arrested by federal officers (Aranda, 2013). The arrest of one member, Arnoldo Rueda-Medin, led to violent retaliation by LFM, including a series of well-coordinated attacks on 16 Federal Police stations. As part of the campaign, a dozen Mexican officers were kidnapped, tortured, and murdered (U.S. Customs and Immigration, 2018). President Calderón responded by sending thousands more Federal Police to Michoacan and the violence escalated.

In 2009, LFM was declared the most dangerous cartel in Mexico based on its cruelty, weapons, and ability to control politicians (Grayson, 2010).

LFM originated in the state of Michoacan. It is present now in several Mexican states and Tierra Caliente. It has become a major problem in the state of Mexico near the capital. Although Michoacan is 600 miles from the U.S. border, it has given rise to criminal organizations of concern to the U.S. LFM, fragmented but survived in a version combating CJNG. Remarkably, so many successful transnational criminals have originated in Michoacan.

Objectives

LFM proclaimed its existence and published its mission statement in newspapers in 2006. In part, it said:

> Who are we? Common workers from the hot lands region in the state of Michoacan, organized by the need to end oppression, the humiliation to which we have constantly been subjected by people who always had power ... Our sole motive is that we love our state and are no longer willing to see our people's dignity trampled on. (De Amicus, 2010, iv)

When LFM tossed severed heads onto a dance floor in 2006, it left an explanatory note saying it only killed those who deserved to die, carrying out di-

---

4   Current Mexican President Sheinbaum commented about the conviction, "The big issue here is how someone who was awarded by United States agencies, who ex-President Calderón said wonderful things about his security secretary, today is prisoner in the United States because it's shown that he was tied to drug trafficking" (CBS, 2024).

vine justice. LFM, it said, did not kill innocent people, women, or for money. LFM communicated with the public by hanging banners and posting messages on social media. It began as a force to counter kidnapping and drug dealers in Michoacana (Grayson, 2010).

The Zetas compared LFM to radical Islamists. A comparison with some merit as both LFM and radical Islamists appeal to religious ideologies to justify their actions. LFM's quasi-Christian fundamentalism was grounded in ideas adopted from John Eldredge's book *Wild at Heart*. Eldredge's basic message was that every man needs a battle to fight, a beauty to rescue and an adventure to live. LFM offered this ideology to its members. LFM had severe penalties for members who transgressed its code of honor by using or selling methamphetamine to local people or who were violent toward women or children. A third infraction of these rules was an executable offense (Grayson, 2011).

Economy/Funding

LFM profits by illegally distributing pharmaceutical products. Heroin was invented by Bayer Corporation to provide a less addictive, more potent pain medication to replace morphine. Both sides in World War II found it expedient to provide methamphetamine for troops, and the stimulant is still prescribed to children with attention deficit disorders. Fentanyl, developed as an analgesic and anesthetic, is 100 times more potent than morphine. Purdue

Pharma, maker of OxyContin, and other corporations indisputably contributed to the opioid epidemic in the U.S.

Cartels did not invent the synthetic products they are trafficking; they capitalized on work done by chemists employed by corporations. The next opportunity for exploitation will be benzimidazole-opioids (aka nitazenes) made in China. These are a class of synthetic opioids up to 800 times more potent than morphine. First synthesized in a laboratory in Switzerland, these compounds were never approved for use in the U.S. Since 2019, there have been over 4,300 reports of nitazenes to the National Forensic Laboratory Information System (DEA, 2024). Nitazenes have been detected already on almost every continent (Inter-American Drug Abuse Control Commission, 2024). Nitazenes are inexpensive to produce but require more sophisticated chemistry than cooking methamphetamine[5] (Zagorski, Myslinski, & Hill, 2020).

Having lost its share of the methamphetamine market in the U.S. to the Sinaloa Cartel and CJNG, a new product would be highly desirable to the NFM. Distributing nitazenes synthesized in China may prove an irresistible marketing opportunity (Dittmar & Rios, 2025). Access to port Lázaro Cárdenas, which LFM fiercely defends from CJNG, allows the Family to swap illegally mined minerals for drugs and chemical precursors from China (Dudley at al., 2024).

LFM was once a major producer of methamphetamine for export to

---

5   See Appendix B for an expert opinion on the cost/ease of synthesis of methamphetamine vs. nitazenes.

the U.S. The new iteration of the cartel is alleged to be involved in the production and distribution of fentanyl. LFM has demonstrated its adaptability. It began synthesizing phenyl-2-propanone (P2P) when international regulations made pseudoephedrine and ephedrine scarce. P2P is more dangerous than earlier methamphetamine. P2P can produce hallucinations, delusions, psychosis and is often laced with fentanyl[6] (Hazelden Betty Ford Foundation, 2011). The number of methamphetamine users doubled in the U.S. between 2009 and 2023 (Vankar, 2024). In any form, methamphetamine is a highly addictive drug that killed over 36,000 in the U.S. in 2023, half as many as those who died of a fentanyl overdose. In California, a gram costs $20 and $40 and a kilo may in the U.S. for $10,500 according to the United Nations. Wholesale, a kilo costs about $600 in Mexico. Producers are looking to expand to more profitable markets. A kilo of methamphetamine can be sold for $20,000 in Europe and $190,000 in Australia or New Zealand (United Nations Office on Drugs and Crime, n.d.). Mexican cartels are wholesalers competing to increase their share of the market (Dittmar, 2024). While some of the stimulant is produced in small clandestine labs, the cartels are investing in industrial-scale facilities and trading finished products for precursor chemicals to avoid creating a money trail (Associated Press, 2024). Industrial-scale laboratories will

facilitate synthesizing nitazenes.

A sense of how production is increasing to meet demand comes by comparing methamphetamine seizures from 2009 to 2025. In 2009, the U.S. Attorney General said Project Coronado "dealt a significant blow to La Familia's supply chain of illegal drugs, weapons and cash flowing between Mexico and the United States" (U.S. Department of Justice, 2009, np). The operation seized 729 pounds of methamphetamine. Two years later, after 20 months of investigation, in an operation called Project Delirium, the DEA seized 635 pounds of methamphetamine from LFM. The Deputy Attorney General said La Familia had been "stripped of its manpower, its deadly product and its profit" (U.S. Department of Justice, 2011). In February 2024, the Mexican Navy seized over 40 tons of methamphetamine from one laboratory (Associated Press, 2024). Cartels appear to be scaling up production. Authorities have won battles, but not the war.

In its early days, LFM opposed the use of narcotics in its home state and made methamphetamine for export only.[7] LFM employed 4,000 natives of Michoacan, paying salaries of $1,500 to $2,000 a month (Grayson, 2011). Just to meet its payroll the organization had to generate millions of dollars a month. The U.S. crack down on methamphetamine had the unintended consequence of creating a business opportunity for LFM (Wilson Center, 2000).

---

6   This is not accidently; opioids depress the central nervous system and methamphetamine speeds it up. Users of this combination may be seeking to balance out the effects.

7   According to a sealed indictment for U.S. District Court, Southern District of NY.

LFM diversified into extortion, kidnapping, human smuggling, selling contraband such as pirated DVDs, loan sharking, illegal mining, and small-scale sales of marijuana and cocaine (Grayson, 2010). Much of its meth-amphetamine business has been taken over by competitors. The U.S. alleges the new version of the cartel is involved in the production and distribution of fentanyl and migrant trafficking. It also aspires to control the lucrative avocado industry.

Effect on the Host State

The general functions of a government are maintaining security, providing public services, and managing the economy. LFM undermined these functions, conducting itself as a de facto state in Michoacana. Poorly paid local police, often incompetent and corrupt, would resign when ordered to do so by LFM and were executed for various "sins." Under Mendez, extensive cooperation occurred between LFM and law enforcement. Investigators found that police permitted members of LFM to use their "radio frequencies and uniforms," and patrol cars were used to block streets to facilitate hitmen getting away (*CE Noticias Financieras*, 2024). In April 2010, 40 men ambushed the armored convoy of public safety chief Minerva Bautista. The attack killed four people and injured nine. By early August, she resigned.

LFM dispensed justice and settled local disputes, including collecting debts for people (for which it took a modest 20 percent of funds collected). In addition to being a well-paying employer, LFM controlled up to 30 percent of official commerce and was connected with 85 percent of legitimate businesses in Michoacan (Wilson Center, 2000). Through extortion, it collected taxes, which were partially redistributed to provide social assistance. Providing public services is an effective strategy for gaining acceptance and providing cover for criminals. All major cartels in Mexico occasionally buy loyalty as a form of public relations, like a legitimate business might sponsor a Ronald McDonald house. LFM went far beyond what was customary, functioning as a parallel structure to the government in providing social services and community development (Fainaru and Booth, 2009). It promoted a faith-based ethos of good works. Like its competitors, it helped the poor during the COVID-19 pandemic, suggesting a sensitivity to its brand identity. Plastic bags of toilet paper, chlorine, canned goods and cereal were distributed to the poor. These bags were labeled with a sticker "Support from La Familia Michoacana,[8] The M Command" and a sticker with a Fish and Strawberry, symbols of the Olascoaga brothers. A video of heavily armed members of LFM passing out provisions was posted on YouTube, ominously in it a cartel member can be heard saying the people should be grateful (Martinez, 2020). Criminal organizations passing out humanitarian aid during a pandemic is direct competition with

---

8   Note they were stilling calling themselves LFM, not NFM.

the state (Bunker and Sullivan, 2020; Flanigan, 2014). La Familia kept crime low in the areas it controlled but did so with a heavy hand. During the pandemic, LFM made rules. It was forbidden to have two sexual partners, no one was allowed to purchase food outside the towns, and masks were mandatory (Martinez, 2023). Perhaps members of LFM were genuinely concerned about people, Mexico had the second-highest case fatality rate in the world from the virus at 4.5 percent (Johns Hopkins, 2023). The President of Mexico said the aid was not helpful and instead the cartel should stop its violence (*Al Jazeera*, 2020).

LFM exploited underlying weaknesses in the political institutions of the state that allowed organized crime and drug trafficking to flourish (Aguirre and Herrera, 2013). In newspaper advertisements, on banners, and over social media, LFM communicated the message that it was protecting the people and would disband once the Mexican government addressed the needs of the people. This is the language of a political insurgency.

Narco Ballads are popular songs glorifying cartel members' success (Garcia, 2006). These ballads are propaganda used in the recruitment of new members. In them, cartel members are portrayed as folk heroes, protecting the poor and battling the rich, demonstrating their ability with brutality (Logan and Sullivan, 2009). The LFM brand presented itself as Pancho Villa for the Twitter generation (Logan and Sullivan, 2009).

LFM established dual sovereignty with the elected government, generating employment, collecting taxes, fulfilling civic functions and keeping order (Grayson, 2010). When LFM fractured into competing groups, KT devised schemes for collecting revenue that people found oppressive. Indigenous people armed themselves to fight illegal logging and the violence of KT.

In Guerrero state, thousands of vigilantes rose up to fight kidnappers and extortionists. The movements spilled over into non-indigenous communities. The vigilante militias that arose in the Tierra Caliente region in February 2013 fought KT. The government failed in its duty to stop the extortion schemes, violating the social contract, and the people took on the task of providing their own security. As a leader of a self-help group said, "when the government is incapable of taking care of the people, of defending the people, the people have the right to defend themselves" (Grillo, 2014). Thousands of vigilantes in 30 municipalities took up arms and successfully waged war against the KT. The Mexican government turned against these vigilantes, arresting many of them (*Al Jazeera*, 2014). Why would the government do this? One Guerrero group proclaimed it was against criminals and bad government who need to see "they are hundreds, we are millions" (Voice of America, 2013).

In theory, the Mexican people have the constitutional right to bear arms in self-defense but only in their homes. Current laws prohibit people from possessing firearms determined to be for military use, thus the people can legally own .22s and single-shot muskets. The Mexican government makes it excessively difficult for citizens to exercise their right to bear arms. The only gun store in the country is on a military base in Mexico City (Cortez, 2021). The paperwork for registering a gun is both extensive and expensive. (The truism that if guns are outlawed only the criminals will have guns seems to apply).

Important founding leaders of the self-defense movement have been killed, including Bruno Plácido. His death came just months after the execution of another vigilante leader Hipólito Mora. These executions wiped out the last of the old guard of the "self-defense" movement (Stevenson, 2023). Avocado wars are giving rise to new groups for self-defense. However, like LFM these groups may evolve into cartels (Arnaud and Jordan, 2023). Cartels directly challenge representatives of the state. On April 4, 2023, a prolonged firefight took place with the police in Michoacán, killing two officers. Cartels in Michoacan employ bomber drones, land mines and Improvised Explosive Devices. The state police in Michoacán now use heavy armored military style trucks (Dittmar and Fernandez, 2025).

The Mexican state itself has been fragmented into a mosaic of opposing state-criminal networks. In 2023, the World Justice Project[9] ranked Mexico 116th out of 142 nations ranked on rule of law, below the Philippines 100, Kenya 101, Niger 109, Sierra Leone 110, and the Russian Federation 113. Corruption at all levels is a major problem in Mexico (Hernandez, 2013). Even Mexico's military has colluded with criminals instead of dismantling organized crime groups (Flannery, 2023). One effect of the battle between the NFM/LFM and CGNJ is the deaths of people running for office and/or their families. In January and February of 2024, 36 people were killed in the run up to elections. By executions, cartels influence the slate of candidates for public office (Berg, 2014). The cartels are now students who graduate from universities for management and accounting services and influencing school curriculum (Felbab-Brown, 2024).

Cooperation with terror groups

The U.S. State Department recently designated several transnational drug trafficking groups as Foreign Terrorist Organizations (FTOs) and Specially Designated Global Terrorists. A number of these organizations have been allies or adversaries of LFM, including CJNG, the Gulf Cartel, and Cartel United. In this sense the cartels cooperate with terror groups by cooperating with one another. In general, it has been deemed unlikely drug trafficking organizations would cooperate with terrorist organizations as such (Moeykens, 2018). If Mexican cartels form alliances with Islamic terrorist groups, it will be for monetary gain. The cartels might collaborate with terrorists to distribute drugs globally. Cartels might facilitate terrorists entering the U.S. on "VIP"

---

9   https://worldjusticeproject.org/rule-of-law-index/downloads/WJPIndex2023.pdf

border crossing packages (Ventura, 2024). Cartels are unlikely to screen customers for connections with Islamic terrorist organizations.

Conclusion

LFM has and continues to operate as a paramilitary criminal insurgency. Vast economic resources and alliances make it a formidable challenge to the Mexican government. Its history of retaliation against government targets is stunning. Mexico has been waging a drug war for decades and experienced more than 460,000 related homicides in the last two (Council on Foreign Relations, 2025). Despite, or perhaps because of U.S. policies, the cartels have grown in number, wealth and ability to corrupt the Mexican government. When President Calderon (2006–2012) declared war on the cartels and captured or killed many leaders, it resulted in fragmenting the cartels into a greater number of more violent groups (Council on Foreign Relations, 2025). LFM survived by fragmenting, regrouping and reemerging. Although the U.S. government has invested more than $1 trillion dollars in the war on drugs, it does not appear to be winning. Pouring money into corrupt nations benefits corrupt officials more than needy citizens (McMullan, 1961). Corrupt governments foster criminal enterprises (Aguirre and Herrera, 2013).

El Salvador's President, Nayib Bukele is the only politician to save a nation from cartel violence and its corrupting influence, but his is a brutal solution to stop Barrio 18 and MS-13 (Mara Salvatrucha). At one time, El Salvador was the murder capital of the world, but now with 1.9 homicides per 100,000 residents, it ranks below Canada's 2.27 and well below Mexico's 24.9 for 2023. Bukele hopes other Latin American nations will adopt his strategy. In his words, "There are many priests, but few are exorcists" (*Time*, 2024).

References

Abi-Habib, M. and Romero, S. (2025, January 22). How labeling cartels 'terrorists' could hurt the U.S. Economy. *New York Times*. https://www.nytimes.com/2025/01/22/world/americas/ mexico-cartel-terrorists-trade.html

Agren, D. (2021, March 18). Mexico ambush: 13 state police killed in attack on convoy. *The Guardian*. https://www.theguardian.com/world/2021/mar/19/mexico-ambush-state-police-killed-attack--convoy

Aguirre, J., and Herrera, H. A. (2013). Institutional weakness and organized crime in Mexico: the case of Michoacan. *Trends in Organized Crime, 16*, 221-238. DOI:10.1007/S12117-013-9197-1

Al Jazeera. (2014, June 28). Mexico arrests well-known vigilante leader. https://www.aljazeera.com/news/2014/6/28/mexico-arrests-well-known-vigilante-leader

Al Jazeera. (2020, April 21). Mexico president tells gangs to stop handing out coronavirus aid. https://www.aljazeera.com/news/2020/4/21/mexico-president-tells-gangs-to-stop-handing-out-coronavirus-aid

Appleby, P. (2023, January 6). 3 Takeaways from the return of the Familia Michoacana. *InSight Crime.* https://insightcrime.org/news/takeaways-return-familia-michoacana/

Appleby, P. (2024, March 5). How criminal groups help expand Mexico's multi-billion-dollar avocado industry. *InSight Crime.* https://insightcrime.org/news/interview/how-criminal-groups-help-expand-mexicos-multi-billion-dollar-avocado-industry/

Aranda, S. M. (2013). Stories of drug trafficking in rural Mexico: Territories, drugs and cartels in Michoacan. *European Review of Latin American and Caribbean Studies/Revista Europea de Estudios Latinoamericanos Y Del Caribe,* (94) 43-66. DOI: 10.18352/erlacs.8393

Arnaud, S.C. and Jordan, T.S. (2023, November 11). Mexico's avocados and arms trafficking: The criminal combination suffocating Michoacan. *El Pais English.* https://pulitzercenter.org/stories/mexicos-avocados-and-arms-trafficking-criminal-combination-suffocating-michoacan

Arrieta, C. (2019, August 9). The story behind two of Mexico's bloodiest cartels. *El Universal. English.* https://www.eluniversal.com.mx/english/story-behind-two-mexicos-bloodiest-cartels/

Associated Press (2024, April 23). Mexico's meth trade has become so lucrative it is exporting the drugs as far away as Hong Kong and Australia. https://fortune.com/2024/04/23/mexico-meth-trade-lucrative-exported-hong-kong-australia/

Associated Press. (2004, February 12). Mexico says it seized over 40 tons of meth from a drug lab in Sonora state. https://apnews.com/article/mexico-biggest-meth-lab-tons-seized-a2130cd0612168553f68d88cf0207d28

Atuesta, L. H., & Pérez-Dávila, Y. S. (2018). Fragmentation and cooperation: the evolution of organized crime in Mexico. *Trends in Organized Crime,* 21(3), 235-261. DOI.org/10.1007/s12117-017-9301-z

Bargent, J. (2016, February 8). 'Nueva Familia' Announces Arrival amid Michoacan Turmoil. *InSight Crime.* https://insightcrime.org/news/brief/nueva-familia-an nounces-arrival-amid-michoacan-turmoil/

BBC. (2016, October 19). U.S. extradites Mexican-Chinese businessman Zhenli Ye Gon. https://www.bbc.com/news/world-latin-america-37699634

Berg, R. (2024, June 21). Mexico's cartel-related violence spikes as elections approach. Foreign Military Studies Office. https://fmso.tradoc.army.mil/2024/mexi cos-cartel-related-violence-spikes-as-elections-approach/

Bunker, R. J. and Sullivan, J. P. (2020, May 8). Mexican cartel strategic note No. 29: An Overview of cartel activities related to COVID-19 humanitarian response. *Small Wars Journal.* https://smallwarsjournal.com/2020/05/08/mexican-cartel-stra tegic-note-no-29-overview-cartel-activities-related-covid-19/

CE Noticias Financieras. (2024, November 24). Cartels linked to detained officials in Mexico State: CJNG, La Familia Michoacana and La Union Tepito. https://www. elimparcial.com/mexico/2024/11/24/los-carteles-ligados-a-los-funcionarios-de tenidos-en-el-estado-de-mexico-cjng-la-familia-michoacana-y-la-union-tepito/

CBS News. (2024, October 16). Mexico's ex-security chief sentenced to nearly 40 years in U.S. prison for taking bribes from cartel run by "El Chapo." https://www. cbsnews.com/news/genaro-garcia-luna-us-prison-mexico-security-chief-car tel-bribes/

Corcoran, A. (2023). La Familia Michoacana (LFM) Mexican drug cartel. In S.N. Romaniuk, M.S. Catino & C. A. Martin (eds.) *The Handbook of Homeland Security* (575-579). CRC Press.

Cortez, D. (2021, April 21). There Can Be Only One: Mexico Has One Gun Store but a Proliferation of Guns. *Michigan State International Law Review.* https:// www.msuilr.org/new-blog/2021/4/21/there-can-be-only-one-mexico-has-one- gun-store-but-a-proliferation-of-guns

Council on Foreign Relations. (2025, February 21). *Mexico's Long War: Drugs, Crime, and the Cartels.* https://www.cfr.org/backgrounder/mexicos-long-war- drugs-crime-and-cartels

Davis, F. G. (2015, December 29). Murder of the founder of 'La Familia' after meet- ing with defense group. *MILENIO.* Translated from Spanish. Asesinan a fundador de 'La Familia' tras reunión con autodefensas. https://www.milenio.com/policia/ pgj-asesinato-fundador-familia-reunion-autodefensas

De Amicis, A. (2010). *Los Zetas and La Familia Michoacana drug trafficking organizations (DTOs).* [Master's thesis, University of Pittsburgh]. https://www.ojp.gov/pdffiles1/234455.pdf

Dittmar, V. and Rios, P. (2025, January 22). How fentanyl producers in Mexico are adapting to a challenging market. *InSight Crime.* https://insightcrime.org/investigations/fentanyl-producers-mexico-adapting-challenging-market/

Dittmar, V. and Fernandez, M. (2025, February 3). Criminal Groups Are Ramping Up Explosives in Mexico. *InSight Crime.* https://insightcrime.org/investigations/fentanyl-producers-mexico-adapting-challenging-market/

Dittmar, V. (2024, April 3). Methamphetamine traffickers in Mexico become global wholesalers. *InSight Crime.* https://insightcrime.org/news/methamphetamine-traffickers-mexico-become-global-wholesalers/

Dudley, S., Dittmar, V., Pforzheimer, A., García, S., Asmann, P., Appleby, P., López, J., Lara, J., Roldan, E., Rico, A.I., Restrepo, J.J., Gaviria, M.I., Portillo, E.L., Guzman, A., Salazar, P. & Storr, S. (2024). The flow of precursor chemicals for synthetic drug production in Mexico. *InSight Crime.* https://insightcrime.org/wp-content/uploads/2023/08/Precursor-Recommendations-InSight-Crime-Ibero-May-2024.pdf

Fainaru, S. and Booth, W. (2009, June 13). A Mexican cartel's swift and grisly climb. *The Washington Post.* https://swap.stanford.edu/was/20090613032245/http://www.washingtonpost.com/

Felbab-Brown, V. (2024, March 14). Mexican Cartels, Fentanyl, and the Global Synthetic Drugs Revolution. Georgetown Americas Institute. https://americas.georgetown.edu/events/mexican-cartels-fentanyl-and-the-global-synthetic-drugs-revolution

Ferri, P. (2022, December 20). The never-ending scourge of La Familia Michoacana. *El Pais, International.* https://english.elpais.com/international/2022-12-20/the-never-ending-scourge-of-la-familia-michoacana.html

Flanigan, S. T. (2014). Motivations and implications of community service provision by La Familia Michoacana / Knights Templar and other Mexican drug cartels. *Journal of Strategic Security, 7*(3), 63-83 DOI: http://dx.doi.org/10.5038/1944-0472.7.3.4

Flannery, N. P. (2023, April 11). Are U.S. avocado buyers financing the cartel conflict in Mexico? *Forbes.* https://www.forbes.com/sites/nathanielparishflannery/

2023/04/11/are-us-avocado-buyers-financing-the-cartel-conflict-in-mexico/

Garcia, M. M. (2006). 'Narcoballads': The psychology and recruitment process of the 'narco.' *Global Crime, 7*(2): 200-213 DOI: 10.1080/17440570601014461

Gibbs, S. (2009, October 22). 'Family values' of Mexican drug gang. *BBC News.* http://news.bbc.co.uk/2/hi/americas/8319924.stm

Global Initiative Against Organized Crime. (2024). Violent and vibrant Mexico's avocado boom and organized crime. https://globalinitiative.net/analysis/mexi cos-avocado-boom-and-organized-crime/#:~:text=Avocado%20production%20 in%20Mexico %20is,the%20fight%20against%20organized%20crime.

Grayson, G. W. (2010). *Mexico: Narco-Violence and a failed state?* Transaction Publishers.

Grayson, G. W. (2011). *La Familia drug cartel: Implications for US-Mexican security.* Didactic Press. Kindle Edition.

Grillo, I. (2011). *El narco: Inside Mexico's criminal insurgency.* Bloomsbury Press.

Grillo, I. (2014). Mexico's vigilante militias rout the Knights Templar drug cartel. *Combating Terrorism Center at West Point.* 7(4). https://ctc.westpoint.edu/mexi cos-vigilante-militias-rout-the-knights-templar-drug-cartel/

Guillen, B. (2023, December 12). Death on a soccer field: La Familia Michoac-ana massacre leaves Texcaltitlan on edge. https://english.elpais.com/internation al/2023-12-12/death-on-a-soccer-field-la-familia-michoacana-massacre-leaves-texcaltitlan-on-edge.html

Hackbarth, K. (2020, June 3). Felipe Calderon's Administration Had Ties to Drug Cartels — And the United States Knew All Along. Jacobin. https://jacobin. com/2020/06/felipe-calderons-administration-had-ties-to-drug-cartels-and-the-united-states-knew-all-along

Hazelden Betty Ford Foundation. (2011). *P2P Meth: The newest product of the meth epidemic, and how we got here.* https://www.hazeldenbettyford.org/articles/ p2p-meth

Henkin, S. (2024). The pits: Violence in Michoacán over control of avocado trade. Tracking cartels infographic series. *National Consortium for the Study of Terror-ism and Responses to Terrorism.* https://www.start.umd.edu/publication/pits-vio lence-michoac-n-over-control-avocado-trade

Hernandez, A. (2013). *Narcoland: The Mexican Drug Lords and Their Godfathers.* Verso.

Hopkins, V. (2011). Former leader: No future for Mexican cartel. *Organized Crime and Corruption Project.* https://www.occrp.org/en/investigation/former-leader-no-future-for-mexican-cartel

Infobae. (2019). The love story and betrayal of "El Mencho's" cousin that gave rise to the Jalisco New Generation Cartel. https://www.infobae.com/america/mexico/2019/05/14/la-historia-de-amor-y-traicion-de-la-prima-de-el-mencho-que-dio-origen-al-cartel-jalisco-nueva-generacion/

InSight Crime. (2024, May 27). Jalisco Cartel New Generation (CJNG). https://insightcrime.org/mexico-organized-crime-news/jalisco-cartel-new-generation/

Inter-American Drug Abuse Control Commission. (2024). The emergence of nitazenes in the Americas. https://www.oas.org/ext/DesktopModules/MVC/OASDnnModules/Views/Item/Download.aspx?type=1&id=1045&lang=1

Jones, N.P. (2018). The strategic implications of the Cartel de Jalisco Nueva Generacion. *Journal of Strategic Security, 11*(1): 19-42 DOI: https://doi.org/10.5038/1944-0472.11.1.1661

Johns Hopkins Coronavirus Resource Center. (2023). Morality Analysis. https://coronavirus.jhu.edu/data/mortality

Kenney, M. (2007). *From Pablo to Osama: Trafficking and Terrorist Networks, Government Bureaucrats, and Competitive Adaptation.* Pennsylvania State University Press

Logan, S., and Sullivan, J. P. (2009). Mexico's "divine justice." *ISN Security Watch,* 17, 1-4. https://www.academia.edu/9203722/Mexico_s_Divine_Justice_

Lomnitz, C. (2019). The ethos and telos of Michoacán's Knights Templar. *Representations,* 147(1), 96-123. DOI.org/10.1525/rep.2019.147.1.96

Martinez, C. (2020, April 24). Trans., "La Familia Michoacana debates journalist – 'We should be applauded' for Covid19 help 'Ungrateful ones, will be killed." *Borderland Beat.* https://www.youtube.com/watch?v=OJAgF_WkNCc

Martinez, J. (2023, December 18). La Familia Michoacan punished even infidelity in Texcaltitlan. Milenio. https://www.milenio.com/policia/la-familia-michoacana-castigaba-hasta-la-infidelidad-en-texcaltitlan

Mayen, B. (2023, December 11). How much are they offering in Mexico and the US for the capture of "El Fresa" and "El Pez," the brothers who control La Familia Michoacana? *Infobae.* https://www.infobae.com/mexico/2023/12/11/cuanto-of-recen-en-mexico-y-eeuu-por-la-captura-de-el-fresa-y-el-pez-los-hermanos-que-controlan-la-familia-michoacana/

McMullan, M. (1961). A theory of corruption. *The Sociological Review, 9*(2), 181-201. DOI.org/10.1111/j.1467-954X.1961.tb0109

Moeykens, J. (2018). *An Assessment of the likelihood: Potential cooperation between Mexican cartels and Al Qaeda or ISIS.* [Master's thesis. Naval Postgraduate School]. https://www.usmcu.edu/Portals/218/CAOCL/files/18Jun_Moeykens_Justin.pdf?ver=2019-04-17-133149-740

Rainsford, C. (2019, September 30). Mexico's cartels fighting it out for control of avocado business. *InSight Crime.* https://insightcrime.org/news/brief/mexico-car-tels-fighting-avocado-business/

Shortell, D. (2024, July 13). As cartels take a stake in 'green gold,' US and Mexico rethink how avocados reach American kitchens. *CNN.* https://www.cnn.com/2024/07/13/americas/avocado-cartel-us-mexico-intl-latam/index.html

Smith, B.T. (2021). *The Dope: The Real History of the Mexican Drug Trade.* W. W. Norton & Company.

Stevenson, M. (2023, October 17). Well-known leader of a civilian 'self-defense' group has been slain in southern Mexico. https://apnews.com/article/mexico-vig ilante-selfdefense-armed-citizens-killed-e14da7fb6ae2383ec8b5994648872987

The Universal. (2022, June 8). Leader of La Familia Michoacana assassinated. https://es-us.noticias.yahoo.com/asesinan-l%C3%ADder-familia-michoacana-060000732.html

Time. (2024, August 29). How Nayib Bukele's 'iron fist' has transformed El Salvador. https://time.com/7015598/nayib-bukeles-iron-fist-el-salvador/

United Nations Office on Drugs and Crime: Drug prices. https://dataunodc.un-.org/dp-drug-prices.

U.S. Customs and Immigration. (2018, April 11). 'La Familia Michoacán' drug cartel leader sentenced to 43 years in federal prison for trafficking thousands of kilograms of methamphetamine into U.S. [Press release]. https://www.ice.gov/news/releases/la-familia-michoacan-drug-cartel-leader-sentenced-43-years-federal-prison-trafficking

U.S. Drug Enforcement Administration. (2024). *Benzimidazole-Opioids.* https://www.deadiversion.usdoj.gov/drug_chem_info/benzimidazole-opioids.pdf

U.S. Department of Justice. (2009, October 22). More than 300 alleged La Familia Cartel members and associates arrested in two-day nationwide takedown. [Press release]. https://www.justice.gov/archives/opa/pr/more-300-alleged-la-familia-cartel-members-and-associates-arrested-two-day-nationwide

U.S. Department of State. (2024, June 20). United States sanctions key La Nueva Familia Michoacana members. [Press release]. https://2021-2025.state.gov/united-states-sanctions-key-la-nueva-familia-michoacana-members/#:~:text=In%20close%20coordination%20with%20the,Busto%20and%20Josue%20Ramirez%20Carrera.

U.S. Department of Treasury. (2022, November 17). Treasury sanctions illicit fentanyl-trafficking La Nueva Familia Michoacana and its leaders. [Press release]. https://home.treasury.gov/news/press-releases/jy1116

Vankar, P. (2024). Number of Americans who used methamphetamine in the past year 2009-2023. *Statista.* https://www.statista.com/statistics/611697/methamphetamine-use-during-past-year-in-the-us/

Ventura, J. (2024, June 24). Mexican cartels offer 'VIP' border crossing packages to migrants. *Border Report.* https://www.borderreport.com/immigration/border-crime/mexican-cartels-offer-vip-border-crossing-packages-to-migrants/

Voice of America. (2013). Aparece nuevo grupo armado en México. https://www.vozdeamerica.com/a/aparece-nuevo-grupo-armado-mexico/1802638.html.

Washington Times (2004, January 12). Prison escape linked to Gulf Cartel. https://www.washingtontimes.com/news/2004/jan/12/20040112-091547-5012r/

Wison Center. (2000). *La Familia Michoacana.* https://www.wilsoncenter.org/article/la-familia-michoacana

Worthman, S. (2011, December 16). The rise of the La Familia Michoacana. *E-International Relations.* https://www.e-ir.info/pdf/15706

Zagorski, C.M, Myslinski, J.M., and Hill, L.G. (2020). Isotonitazene as a contaminant of concern in the illegal opioid supply: A practical synthesis and cost perspective. *International Journal of Drug Policy*, 86: 102939. DOI: 10.1016/j.drug po.2020.102939

APPENDIX A

An expert opinion by Dr. Rhys Thomas[10] as to whether it is easier/cheaper to synthesize methamphetamine or nitazenes, in his words:

The central part of the nitazene molecules comes from a Chinese patent that describes the process as cheap and easy. However, it involves several steps with dangerous reagents like concentrated sulfuric acid and concentrated nitric acid. The resulting compound is a "common" building block for protein synthesis. But significant skill is required, and the starting materials are regulated. However, most of the leftovers would be fairly easy to disperse into the environment with a low risk of getting caught. This is in contrast to meth, which a trained nose can track just by driving down the street with the windows open. Also, nitrazine synthesis is less likely to explode than meth synthesis. The last step in nitazene synthesis requires chloroform, which is harder to hide, but at least is not flammable.

As I understand meth synthesis, small "labs" produce it in varying purity. Sophisticated equipment is not required. Things are heated, but generally cooling is not required. Nitrazine synthesis, in contrast, needs much more sophisticated glassware and multiple steps that must be performed at cold temperatures, which requires a knowledge of mixing theory and thermodynamics (not a big deal for a synthetic chemist, but perhaps out of reach for a druggy). So, the synthesis would need to be pushed up the food chain, increasing liability for those near the top of the cartel. With meth synthesis, they can push the liability downstream. Those little meth labs have a tendency to be discovered or to blow up if they stay in the same place too long. The people at the top of the cartels absorb these losses without concern. Another "cooker" is easy to find.

If cartels wanted to get into the nitazene business, surely they could circumvent the regulations surrounding obtaining the starting materials, which are cheap and already exist in sufficient quantities. I suspect that they could find a ready supply in Southeast Asia, maybe even in India. Corruption is rife in both regions. Also available are drug-synthesis facilities that would divert some equipment on the night shift (when they usually make the drugs that are shipped to Africa, but that profit margin is not very good). That would make the cost of production very low with predictable potency. The synthetic methods already exist and are used regularly for large-scale pharmaceutical synthesis. The only drawback I see is that they would be easier to find and the loss of one production facility would be significant.

---

10 Dr. Thomas is a professional chemist who did graduate work at Oak Ridge National Laboratories, contract work for the government, and founded Fayette Environmental Services. He is also a friend who found this an interesting question.

APPENDIX B

Gibbs, S. (2009). 'Family Values' of Mexico drug gang. BBC News

MEXICAN DRUG CARTELS' MAIN AREAS OF INFLUENCE

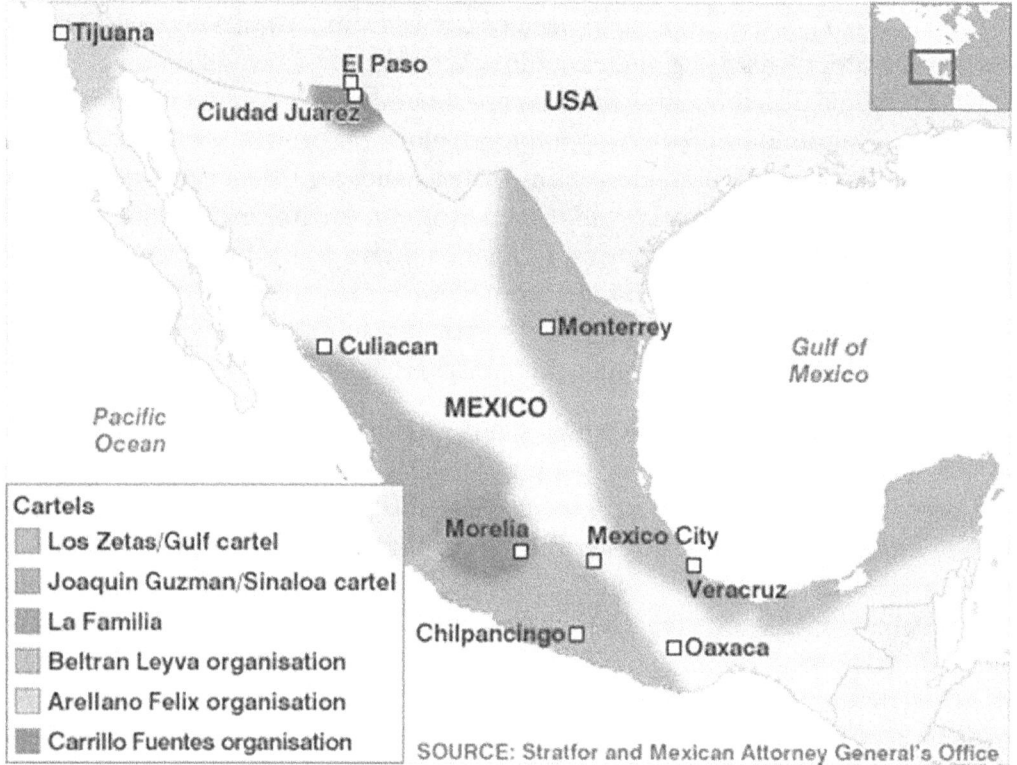

Tijuana

El Paso

Ciudad Juarez

USA

Culiacan

Monterrey

Gulf of
Mexico

Pacific
Ocean

MEXICO

Cartels

Los Zetas/Gulf cartel

Joaquin Guzman/Sinaloa cartel

La Familia

Beltran Leyva organisation

Arellano Felix organisation

Carrillo Fuentes organisation

Morelia

Mexico City

Veracruz

Chilpancingo

Oaxaca

SOURCE: Stratfor and Mexican Attorney General's Office

Security and Intelligence • Volume 10, Number 1 • Spring 2025

Justification (Justified True Belief) in the Intelligence Context: Testimony and Justification Defeaters

Linda Johansson[1]

Abstract

In an argument map arguments are structured in reasons or evidence and premises, providing oversight of the information at hand. The step after constructing an argument tree is to *evaluate* the evidence or reasons for or against the main claim in terms of relevance and credibility. To deal with the different contents in the argument tree, it is necessary to have an extended argument tree where the different types of content might need different types of justification. It is also, as will be shown in the following, important to expand the tree and explore so called *justification defeaters* in the intelligence context, particularly connected to testimony and the component of agency (or lack of it), with the risk of deliberate deception or gullibility in either the one giving the testimony or the one receiving it. One suggestion is to add boxes and use a list of justificatory defeaters for boxes of testimony in the argument tree. The question this paper aims to answer is: How can the philosophical notion of "justified belief" (as a general understanding of justification) be understood and used in the intelligence context connected to testimony and evaluating sources? The suggestion is: as a belief or statement where justification defeaters have been taken into account explicitly.

Keywords: argument mapping; epistemology; justification; testimony; justification defeaters

Justificación (Creencia Verdadera Justificada) en el Contexto de Inteligencia: Testimonio y Derrotadores de la Justificación

Resumen

En un mapa de argumentos, los argumentos se estructuran en razones o evidencias y premisas, lo que permite supervisar la información disponible. Tras construir un árbol de argumentos, el paso

---

1   Linda.Johansson@fhs.se

doi: 10.18278/si.10.1.4

siguiente consiste en evaluar la evidencia o las razones a favor o en contra de la afirmación principal en términos de relevancia y credibilidad. Para abordar los diferentes contenidos del árbol de argumentos, es necesario contar con un árbol de argumentos extendido donde los diferentes tipos de contenido puedan requerir diferentes tipos de justificación. También es importante, como se mostrará a continuación, expandir el árbol y explorar los denominados factores de justificación en el contexto de inteligencia, particularmente en relación con el testimonio y el componente de agencia (o su ausencia), con el riesgo de engaño deliberado o credulidad tanto en quien da el testimonio como en quien lo recibe. Una sugerencia es añadir casillas y utilizar una lista de factores de justificación para las casillas de testimonio en el árbol de argumentos. La pregunta que este artículo pretende responder es: ¿Cómo se puede entender y utilizar la noción filosófica de «creencia justificada» (como una interpretación general de la justificación) en el contexto de inteligencia, en relación con el testimonio y la evaluación de fuentes? La sugerencia es: como una creencia o afirmación en la que se han tenido en cuenta explícitamente los factores que la justifican.

Palabras clave: mapeo de argumentos; epistemología; justificación; testimonio; derrotadores de la justificación

情报语境中的证成（确证的真实信念）——证词与证成破坏因素

摘要

在论证图中，论证由理由或证据和前提构成，用于对现有信息进行梳理。建构论证树的下一步是从相关性和可信度的角度评价那些支持或反对主要主张的证据或理由。为了应对论证树中的不同内容，有必要建构一个扩展的论证树，其中不同类型的内容可能需要不同类型的证成。正如下文所示，扩展论证树并探究情报情境中所谓的"证成破坏因素"（justification defeaters）也至关重要，尤其是与证词和主体性（或主体性缺失）相关的证成破坏因素，因为提供证词或接受证词的人都可能存在故意欺骗或轻信的风险。建议在论证树中添加方框，并在证词方框中使用证成破坏因素列表。本文旨在回答的问题是：在与证词和评价信息来源相关的情报语境中，如何理解和运用"确证信念"（作为对证成的一般理解）这一哲学概念？建议则是：将其作为一种信念或陈述，其中证成破坏因素已被明确考虑。

关键词：论证图，认识论，证成，证词，证成破坏因素

Introduction

The analysis of intelligence involves critical thinking, reasoning and making proper or valid inferences (e.g., Clark, 2020; Hendrickson, 2018). One method for systemizing critical, logical thinking in a transparent way, is argument mapping. In an argument map, arguments are structured in reasons or evidence and premises, placed in boxes below a main claim, which also can be referred to as a thesis or hypothesis (Johansson, 2024). By doing this, it is easier to gain oversight of the information at hand. There are also suggestions that it is important to separate different kinds of reasons or evidence in the boxes of an argument map: facts, tentative facts, judgments, and value judgments (Johansson, 2024). It is also suggested that the argument map should be as clear as possible and instead of numbers and likelihoods we should use the notion of justified belief.

The next step, after constructing an argument tree, is to *evaluate* the evidence or reasons to support or disconfirm the main claim in terms of relevance and credibility. Solely relying on intuition, however, might be problematic. Intuition is significantly prone to error, which is one of the most important arguments for using structured analytical methods (Whitesmith 2022b, 835).

Whitesmith mentions justification: "By making the justification for our judgments explicit, errors in judgment become more visible" (Whitesmith 2022b, 835). She argues that intelligence analysis fundamentally is an "exercise in epistemic justification" (Whitesmith 2022b, 838), and further points out that "guidance on what valid justification for beliefs includes, and how much weight different kinds of justification can carry in the reduction of uncertainty in intelligence is an underdeveloped area of academic research and might benefit greatly by looking at the philosophical area of epistemology, and particularly the criterion of justification – justified true belief (JTB)-theory" (Whitesmith 2022b, 835-6). JTB "seeks to identify general standards for determining a sufficient degree of justification for true beliefs that can be applied to all types of sources of information and all types of mechanisms through which knowledge can be gained. The focus on a sufficient degree, recognizes the near impossible task of identifying pure knowledge" (Whitesmith 2022b, 836).

When it comes to the task of evaluating or weighing evidence, the vital task would be to determine credibility in terms of completeness and soundness (Clark 2020, 180). In terms of professional practice, it is sometimes considered best practice to lead off and conclude/transition in the analyst's own thoughts.[2]

The notion of justified belief in the intelligence context, can be used for such evaluation. The idea of justified belief stems from the third criterion in the classic definition of knowledge.[3] The notions of *truth* and *belief* in the JTB-theories can and have been questioned in the philosophical debate. In this context it is more useful to talk about justification in broader terms and leave issues on truth and belief aside. The question is how philosophical ideas on justification can be transferred and used in intelligence analysis, that is, how the notion of justification in epistemology can help evaluate arguments/evidence in the boxes of an argument map in intelligence analysis.

Considering the different contents in the boxes of an argument tree, there might be different criteria for when a belief is justified. One distinction is between tangible evidence and testimonial evidence. Testimonial evidence is central for intelligence, and the most difficult when it comes to justification since it is prone to error in other ways than tangible evidence. This article will focus on justified belief connected to testimony.

To deal with the different contents, it is necessary to have an extended argument tree where the different types of content might need different types of justification (not necessarily connected to foundationalism and coherentism, mentioned below). It is also, as will be shown in the following, important to expand the tree and explore so called *justification defeaters* (or epistemic defeaters) in the intelligence context, particularly connected to testimony and the component of agency (or lack of it), with the risk of deliberate deception or gullibility in either the one giving the testimony or the one receiving it. One suggestion is to add boxes and use a list of justificatory defeaters for boxes of testimony in the argument tree.

The question this paper aims to answer is: How can the philosophical notion of "justified belief" (as a general understanding of justification) be understood and used in the intelligence context connected to testimony and evaluating sources? The suggestion is: as a belief or statement where justification defeaters have been taken into account explicitly.

Background

The main categories regarding justification in philosophy (epistemology) are *foundationalism* and *coherentism*. According to foundationalism justified beliefs are based on foun-

2   Thanks to an anonymous reviewer for this information.

3   There are several theories on justification in epistemology. First of all: justification is part of the criteria for knowing something. The classical definition of knowledge is that it consists of *justified true belief*, which can be described like this: Luke knows p "if and only if" (meaning that the conditions that follows are sufficient and necessary – iff for short) (i) Luke *believes* that p, (ii) p is *true*, and (iii) Luke is *justified* in believing p. That is, (iii) is the criterion on justification (Steup and Neta, 2020).

dations—basic beliefs, and other beliefs rest on those basic beliefs. According to coherentism knowledge and justification are structured like a web and there is "truth" and support within the system. While foundationalism and coherentism have different views on truth, both concepts are useful in framing an understanding of underlying truths and knowledge (Steup and Neta 2020).

Steup and Neta argue that for true beliefs to count as knowledge, it is necessary that they "originate in sources we have good reason to consider reliable" (Steup and Neta, 2020). Examples of such sources would be *perception, introspection, memory,*[4] *reason,*[5] and *testimony*. As already mentioned, this article will focus on testimony.

Tecuci et al. discusses *tangible evidence*: this can be objects of various kinds, or sensor records like those obtained by signals intelligence (SIGINT), imagery intelligence (IMINT), measurement and signature intelligence (MASINT) and other possible sources (Tecuci et al., 2016, 3; Johansson, 2024). Testimonial evidence, obtained from human sources, would reside within human intelligence (HUMINT). Tucker argues that from the perspective of epistemology and its five sources of knowledge, HUMINT, COMINT (communications intelligence) and OSINT (open-source intelligence) would be within the realm of the epistemology of testimony (Tucker, 2023, 2).

Tecuci et al. point out that the origin of one of the greatest challenges in assessing the believability of evidence is that different questions must be posed regarding the sources of tangible evidence "than those we ask about the sources of testimonial evidence" (Tecuci et al., 2016, 3.) Of course, it is possible, for instance, to deceive in different ways; it is possible to deceive by producing fake evidence like mock targets or signaling, but there is a difference when it comes to testimony—when a human source claims something, perhaps providing someone with fake evidence, or in turn is being deceived by fake evidence. But there is agency connected to tangible evidence as well, indicating the importance of detecting deception—along with gullibility—in human sources. Tecuci et al. point out

---

4   With our *memory* we can retain knowledge already required, but the problem is that our memory is fallible. Steup and Neta suggest that we distinguish between the following: remembering that *p* (which entails the truth of *p*) and *seeming* to remember that *p* (which does not entail the truth of *p*). Here the question is whether one has reason to believe that one's memory is reliable – or other people's memory, in a testimony. This makes memory and testimony connected, and they need to be explored together. (Steup and Neta, 2020; Johansson, 2024).

5   Regarding *reason*, there is something called *a priori* – that justification can be *a priori*, and this means that one is justified in believing something without any experience. Examples of a priori knowledge would be conceptual truths (such as "All bachelors are unmarried"), and truths of mathematics, geometry and logic (Steup and Neta, 2020). Justification and knowledge are called *a posteriori* or empirical if it is not a priori. Some argue that there is no *a priori* knowledge, that all such knowledge is in fact empirical (Steup and Neta, 2020) but we do not need to go deeper into that here, but for our purposes recognize conceptual, mathematical and logical knowledge a priori (Johansson, 2024).

that with tangible evidence, "believability and credibility may be considered equivalent terms," but with human sources there is a characteristic besides credibility: competence (Tecuci et al., 2016, 4). This is something that cannot be overlooked.

Whitesmith discusses evidentialism, which is an externalist form of foundationalism. It has to do with assessing the credibility of individual pieces of information (evidence) and determining whether a belief is true or false. She argues that the challenge is to find a middle ground between the ideals of evidentialism and the reality of the intelligence context, since this context contains unique challenges and often limited options for independent corroboration like "poor quality and incomplete evidence basis, deliberate attempts to conceal or provide false information," (Whitesmith, 2022, 842). In the next section there will be a description of some attempts to evaluate sources in the intelligence context.

Evaluating Sources

Suggestions for evaluating human sources can be found in NATO's Standardization Agreement (STANAG) 2511 (2003), which is an improvement of the so called Admirality Code. In STANAG there are explicitly based assessments of the reliability of sources on truth frequence in (a relevant class) of past reports. Reliability is graded between A (completely reliable tried and trusted source that can be depended on with confidence) and E (unreliable source which has been used

in the past and has proved unworthy of any confidence) and F (reliability cannot be judged—a source which has not been used in the past). The ambiguity of these criteria has been criticized (Capet and Revault-D'Allonnes, 2014, 112-125; Tucker, 2023, 3). Another critique is how to deal with sources that have similar total evaluations like B3 and C2. As I will argue, it is necessary to specify the reasons for evaluations, and to use justification defeaters.

STANAG has also been criticized for failing to distinguish the "genealogies of information, for example whether they are primary or secondary sources; and the competence of the sources to make the kind of judgments they convey" (Tucker, 2023, 4). STANAG has explicated information credibility as "priors," which means the extent to which new information is coherent with previous analysis (Tucker, 2023, 4).

"Yet, if it can be stated with certainty that the reported information originates from another source than the already existing information on the same subject, it is classified as "confirmed by other sources" and is rated "1"" (Irwin and Mandel, 2019, 505). 2 rates what is probably true, "if the independence of the source of any item or information cannot be guaranteed, but if, from the quantity and quality of previous reports its likelihood is nevertheless regarded as sufficiently established" (Irwin and Mandel, 2019). 3–5 ratings are not related to independent sources but are strictly about prior probabilities as coherence with what has already been established." (Tucker, 2023, 4)

Tucker argues that even though many texts about HUMINT methodologies sometimes mention "explicitly or imply implicitly basic Bayesian concepts such as likelihood, prior, expectedness, and posterior probability, they do not always distinguish them clearly or correctly conceptualise their relations" (Tucker, 2023, 2). This is a problem for the evaluation of human sources.

The Admirality Code or NATO system separates (a) the reliability of sources from (b) the credibility of the information they transmit (Tucker, 2023, 3-4). Source reliability is graded from A (completely reliable) to E (F), and credibility from 1 (completely credible) to 5 (6). (Cf. Headquarters, Department of the Army, 2006, Appendix B). The criteria has been criticized for being imprecise—statements like "usually reliable" can be interpreted differently (Tucker, 2023, 4; Irwin and Mandel, 2019, 505).

Another alternative are the French intelligence grades, where "quality" and "content" convey similar, but not identical meanings (Tucker, 2023, 3; Capet and Revault-D'Allonnes, 2014, 108). Tucker points out that Tecuci, Schum, Marcu, and Boicu (2016, 122–124) suggested that analysts evaluate the believability of testimonial evidence "by breaking it into competence that depends on the access of the sources to the information they purportedly transmit, understandability-their ability to perceive the information they receive and interpret it correctly, and credibility which they analyse as a combination of veracity, objectivity, and observational sensitivity" (Tucker, 2023, 3). Tucker believes that Tecuci et al.'s concepts of believability, competence and credibility partly overlap with Bayesian reliability, as stated explicitly in Schum and Morris (2007), and continues:

> "But reliability may be affected by other unmentioned vectors applicable only to particular or few testimonies. Understandability is unnecessary because sources who do not understand or believe in their own testimonies can still transmit information. For example, a mole may ask an innocent colleague to pretend to be a double agent to misinform the enemy, while actually giving that colleague true and valuable information to pass to the other side." (Tucker, 2023, 3)

Tecuci et al. suggest the following approach to deal with testimonial sources: In order to discern how believable testimonial evidence is, one should break it down into (a) competence that depends on the access of the sources to the information they purportedly transmit, (b) understandability: their ability to perceive the information they receive and interpret it correctly, and (c) credibility: a combination of veracity, objectivity and observational sensitivity (Tecuci et al., 2016, 122-124).

As I will argue, it is necessary to specify the reasons for evaluations further, and to use justification defeaters.

Justification and Testimony in the Intelligence Context

In order to demonstrate the need for justification defeaters connected to testimony, the "Curveball-case" as described by James B. Bruce (2008) will be used. Curveball was the codename for a human source claiming to work as a chemical engineer in Iraq's weapon of mass destruction (WMD) program and whose intelligence could be found in "roughly a hundred detailed reports" used to build a rationale for invading Iraq in 2003. Curveball was later exposed as a fabricator.

The essence of the case has been placed into argument trees. We will look at some suggestions from Bruce, Tucker, and Whitesmith regarding evaluation of testimony from sources connected to these trees, beginning with Bruce.

Bruce describes four principal ways of acquiring knowledge: *authority*, *habit of thought* (prejudice in individuals and conventional wisdom in groups; Bruce, 2008, 173), *rationalism*, and *empiricism* (Bruce, 2008, 172-177).

Regarding the problem with relying on authority, Bruces argues that "users cannot easily access any antecedent epistemologies by which the knowledge was created and therefore cannot assess the veracity or sources of possible error" (Bruce, 2008, 73). Epistemologies should probably be understood as was of acquiring knowledge.

In 2002, George W. Bush and senior administration officials claimed to know that Iraq possessed a major program om weapons of mass destruction (WMD) (Argument from authority).

The question is how to come up with counter arguments such as C1P1— how to be able to place such a statement in a tree, since the fact that Curveball was exposed as a fabricator was not known at the time. This is a matter of scrutinizing knowledge from authority, and where the use of justification defeaters might help. It seems necessary to evaluate each box and forcing oneself to come up with counter arguments, but it should be done in a systematic manner. Evaluating each box separately and finding counter arguments might help to see what more information should be acquired.

Bruce points out that even lacking solid information, analysts concluded that Iraq must have had the weapons even if they were not seeing them. "In fact, not seeing them seemed to provide evidence that Iraq had them." It is important to realize that logic is no better than the content of its premises, as Bruce points out: "If the premises contain error, even the soundest reasoning can only reproduce error" (Bruce, 2008, 182).

Main claim: **Iraq possesses CV (chemical weapons)**

P1 (first pro argument for main claim): (NIE): Iraq has an active CV program involving the production of mustard, sarin, cyclosarin and VX and has stockpiled 100-500 metric tons

P1P1 (first pro argument for P1): Images of Samarra type tanker trucks. (Bruce: simplified, fully elaborated in WHO commission report, 122-124).

P1P1P1 (first pro argument for P1P1): Imagery analysts say so (authority – a combination of empirical observations and expert judgment).

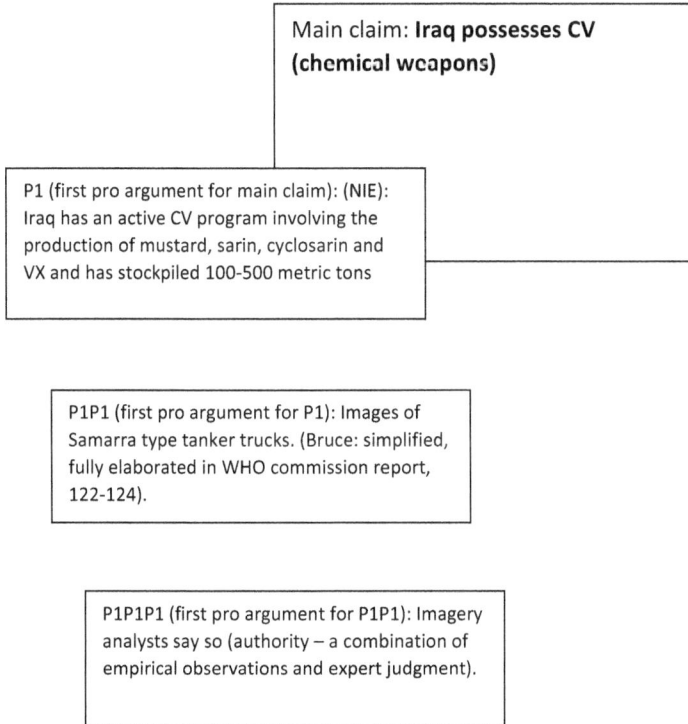

ARGUMENT TREE 2

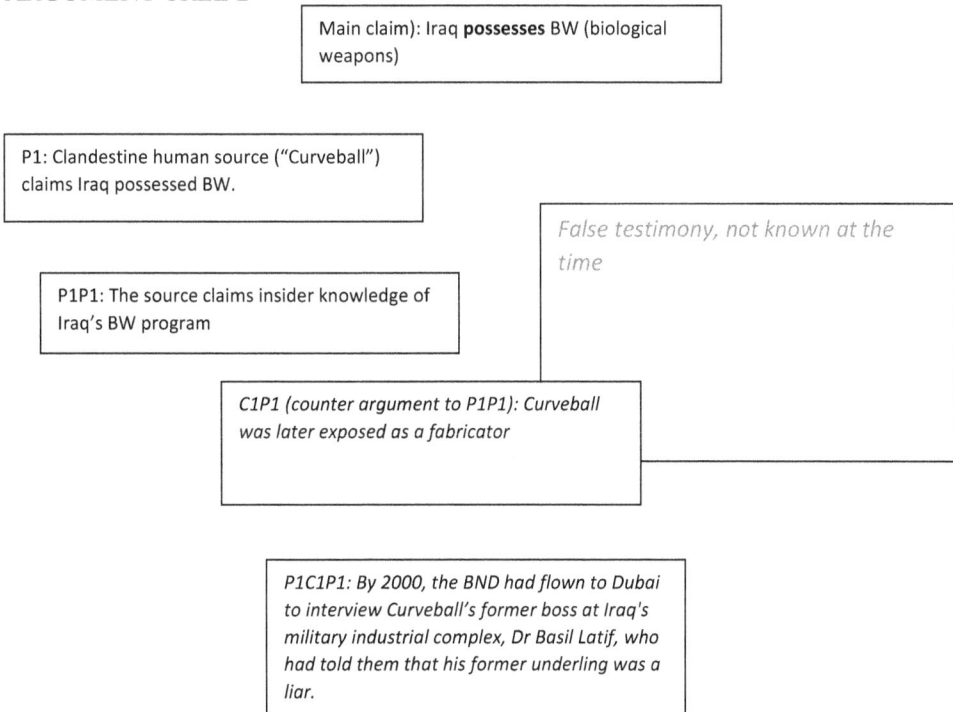

Main claim): Iraq **possesses** BW (biological weapons)

P1: Clandestine human source ("Curveball") claims Iraq possessed BW.

False testimony, not known at the time

P1P1: The source claims insider knowledge of Iraq's BW program

C1P1 (counter argument to P1P1): Curveball was later exposed as a fabricator

P1C1P1: By 2000, the BND had flown to Dubai to interview Curveball's former boss at Iraq's military industrial complex, Dr Basil Latif, who had told them that his former underling was a liar.

Empirical observation seems to have played "a startingly minimal role" (Bruce, 2008, 182) in the NIE on Iraq's WMD: The IC had no direct evidence of WMD in Iraq at the time the estimate confidently asserted knowledge of Iraq's weapon systems.

Bruce describes that what little observable evidence there was of CW, BW and nuclear reconstitution was not only over interpreted but also was not assessed relative to any available evidence to the contrary: "The senior defector's reporting in 1995 that Saddam had shut down the WMO programs four years earlier was simply disregarded" (Bruce, 2008, 182).[6]

Bruce argues that the NIE was an epistemological "perfect storm" because all four ways of acquiring knowledge failed in terms of producing reliable knowledge: authority (faulty), habit of thought (unquestioned), reasoning (flawed), and empiricism (nearly absent). He suggests that the errors that each method produced are "expected outcomes of epistemologies whose strengths do not extend to discovering and correcting their own errors. For more reliable analysis, we need to consider what a more scientific approach might offer" (Bruce, 2008, 182). He suggests using the following insights from science: the use of hypotheses, objective methods, transparency, replicability, peer review, and provisional results.

The first, hypotheses, is considered one of the most important differences between scientific and non-scientific activity, but he points out that hypothesis-testing with statistical tools used for quantitative data is relevant for only a very small number of problems facing the intelligence analyst, and most of them are qualitative (Bruce, 2008, 183). He argues that ACH, as described by Heuer or Whitesmith, offers potential for testing qualitative hypotheses, and there is software for computed aided analysis, for collaborative analysis, and a Bayesian approach, and that this might take care of the requirements of objectivity, transparency, replicability which are vital self-corrective mechanisms.). Bruce argues that the social sciences need to develop qualitative hypothesis-testing techniques for intelligence analysis. Until then, he argues that ACH offers the best technique. That might be true, but the ACH is something that happens before an argument tree is constructed. And it does not quite help with the problems viewed above.

Bruce suggests that one can use the self-corrective mechanisms from science—that the attributes that provide science with self-corrective mechanisms are still largely within reach (Bruce, 2008, 184). An example of a self-corrective mechanism would be to actively use or spell out justification defeaters, which we will look at later on. That is, to use justification defeaters

---

6   This was reporting of Hussein Kamil, SH's son-in-law, who defected in 1995. WMD Commission report, 52; ISG Report 46, and the Butler Report, formally Committee of Privy Counselors, Review of Intelligence on weapons of mass destruction (London; Stationary Office, 2004, 47-48, 51).

from epistemology as a self-corrective mechanism.

Bruce suggests evaluation based on the four ways. First, authority (the basis of the authority), second habit of thought, third analytical logic (which is only as good as the premises the conclusion rests on), and fourth making sure that observable data is true.

The first and second can be dealt with by looking at justification defeaters. The third can be dealt with by turning the argument tree upside down, letting the boxes with pro- and contra arguments be premises and the main claim a conclusion. The fourth has to do with tangible evidence and the evaluation of that.

Tucker talks about frequencies connected to evaluation of human sources and points out that many (Thagard, 2005; Plantiga, 1993, 78–82; BonJour, 2010, 155) "have argued against the so called frequentist interpretation of reliability because often it is impossible to calculate statistically meaningful frequencies of true statements in a set of claims when the class is too small" (Tucker, 2023, 3). Tucker also points out that "even when there is sufficient evidence for estimating frequencies, an infrequent or even just one, but 'big' lie may plunge the reliability of a source, even if the overall frequency of true testimonies by this witness remains high" (Tucker, 2023, 4). In the context of intelligence gathering, assuming or using frequencies to evaluate reliabilities "may be used by an adversary to mislead by bombarding analysts with verifiably true but low

value intelligence from a source, before dropping into the information stream a single highly misleading and harmful falsehood" (Tucker, 2023, 4).

Tucker's suggestion is to use reliability, coherence and independence as central notions, and he has a three stage Bayesian modular model of generation of knowledge to acquire knowledge, also from testimonies of low or indeterminate reliability, if they are coherent and independent, and the prior probability of the information they generate is low. He argues that this modular form of inference is useful for restructuring HUMINT, OSINT and SIGINT analysis, as a vantage point for criticizing current practices and as analytic framework for analyzing mistakes in intelligence analysis. Reliability: "the ratio of information preserved at the end of the transmission to the one sent at its beginning." Coherence: "positive correlation between information signals." Independence: "the absence of information transmission between the sources."

Tucker argues that explicit Bayesian conceptualisation might clarify and disambiguate concepts used by analysts. That would not, however, help with the evaluation of boxes with testimonial content in an argument tree. It is unclear how his suggestions would have prevented the problems with the Curveball-tree.

Whitesmith discusses justified true belief-theory (JTB) and points out the difficulties in implementing such standards in an imperfect reality inevitably, outlines the challenges and and suggests pragmatic solutions for how they can be overcome. But here the most important thing is justification.

Whitesmith discusses reliabilism (an externalist version of foundation-alism), which she considers the most well-developed JTB-theory.[7] The short description is that one is justified in be-lieving a proposition to be true if one's belief was acquired by a reliable meth-od. In this context, it would be that the content of a box is justified if acquired by a reliable method. But what about testimony? One possible reliable meth-od would be the argument tree with ad-ditional justification defeaters.

She argues that process reliabi-lism is too vague. One might wonder what a process being reliable actually entails, and to what degree it is reliable. She points out that the term cognitive process refers to "any mental process in-volved in gaining knowledge and com-prehension. This encompasses a wide range of higher-level brain functions, including, for example, thinking, imag-ination, perception, problem-solving, language, memory and judgment. The mental processes to which process re-liabilism refers face a significant degree of individual difference. No two brains are the same" (Whitesmith, 2022, 838). One person's memory may be more re-liable than another, for example. And it is not open to external observation.

A suggestion for dealing with the difficulties connected to this might be to measure how often analysts turn out to be correct in their judgements—to confer a truth ratio to their analytical capabilities. This approach would be an application of verification of truth and has been incorporated into some versions of process reliabilism. White-smith argues that such an approach would place analysts in an unfairly pressured work environment and that it would also be theoretically invalid from an epistemological perspective. Correct predictions do not necessarily entail that they were formed by reliable cog-nitive processes—being right does not mean that your beliefs were justified.

Whitesmith points out one way to solve the problem—to look at how often analysts are correct in their judg-ments, but that does not mean that those judgments actually were formed by reliable cognitive processes, it might be a question of luck or influence from someone who sometimes provides cor-rect information, sometimes not. "Being right does not mean that your beliefs were justified" (Whitesmith, 2022, 838). She also discusses the problem regard-ing analysts with seemingly similar cog-nitive processes in terms of the quality of the process, and the fact that they might come to different conclusions even though the conclusions are based on the same information, and that it in such cases might be difficult to decide who has greater epistemic justification.

Whitesmith suggests using the key principles of indefeasibilism, evi-dentialism, and processreliabilism for ways to improve intelligence analysis best practices and provides four recom-mendations.

---

7   First developed by Alvin Goldman (a reliabilist theory of knowledge).

First, making better use of formal logic as a way of "externalizing and verifying that intelligence assessments are formed through a reliable cognitive process" (Whitesmith, 2022, 844). This can be facilitated by an argument tree, particularly if turned upside down, so the main claim is a conclusion at the bottom, following from the premises (reasons in the boxes) above, as mentioned previously.

Second, she suggests ensuring that multiple logically valid hypotheses are taken into consideration in forming intelligence assessments, and then there are some basic improvements to guidance on using the original version of ACH that JTB would advocate: "Most current versions of ACH either do not include credibility of information as a ranking system in the ACH process, or do not incorporate credibility of information scores into the overall scoring system. In this capacity, most versions of ACH do not adequately take epistemic justification into account" (Whitesmith, 2022, 844). Whitesmith also suggests that the use of ACH in intelligence communities should be on the basis that the recommended versions of the technique necessarily include two ranking systems: "one that assesses the evidential quality (credibility) of information and one that assesses the degree to which the information supports the truth value of the hypotheses" (Whitesmith, 2022, 844). The question is how to accomplish this.

She suggests that formal logic should be incorporated into the ACH design and that hypotheses included in the technique should first be vetted for logical validity, which requires competing hypotheses to be broken down into their underlying premises (Whitesmith, 2022, 844). This would imply an argument tree. Whitesmith points out the need for reformulating a version of ACH with these suggestions offers a pragmatic way to align intelligence analysis with the ideals enshrined in JTB theory for how best to establish epistemic justification for beliefs. However, to be able to provide the best alignment with these principles, this approach "needs to be paired with adequate guidance on how to judge the degree of evidential epistemic justification of source information. Arguably the greatest challenge facing the development of new methodological approaches is the current lack of appropriate guidance provided in ACH manuals or training and guidance documents as to what methods or standards should be used to assess the credibility of information" (Whitesmith, 2022, 845).

Third, ensure that "the principle of eliminative induction is followed in attempting to seen intelligence information to compare against equally valid hypotheses when existing information provides an equal degree of evidential epistemic justification" (Whitesmith, 2022, 845).

Fourth, ensure that "guidance on judging the epistemic justification of intelligence information is of the highest theoretical quality" (Whitesmith, 2022, 845).

Bruce and Whitesmith suggest using ideas from science, as shown, but the question is how to *actually* make this happen in a systematic manner in order to deal with the problems in the argument tree looked at previously. As mentioned, it is also unclear how Tucker's suggestions would help with the problem in the Curveball-tree, that is, to find the problematic boxes with information not known at the time. We will now look at the suggestion from justification defeaters.

Testimony According to Philosophy

*T*estimony differs from the sources which were mentioned earlier—it lacks a cognitive fallacy of its own. We know p because someone said p. It has been argued by Steup and Neta that this should be understood widely, so that it might include talk and writings in blogs, articles, television, radio, and other media (Johansson, 2024).

There are different approaches to consider. For instance, the so-called *track record approach*. It is argued that we have a tendency to accept testimonial sources as reliable, thinking that they are credibly, as long as there is no reason to believe otherwise (Steup and Neta, 2020; Johansson, 2024). It might be the case that a sign of reliability is that there is a long track record (Fricker, 1994, 2007). Another view in philosophy is that one is prima facie justified to trust a testimonial source.[8] This would not be appropriate for the intelligence

context where there is a lot of potential deception involved.

To acquire knowledge is difficult enough in normal circumstances, but in the context of intelligence, you have to deal with phenomena like "deception, misinformation, the fabrication of intelligence for financial gain, and active attempts by adversaries to keep information hidden" (Whitesmith, 2022a, 837). This is why it is extra important to have boxes with justification defeaters connected to boxes whose contents are based on testimony.

What is needed in the intelligence context is a way to determine to what extent a testimonial source can be trusted, not just simply accept it, since that might lead to devastating results. It can be argued that many mistakes in the intelligence history actually stem from acting as if one would be prima facie justified to trust certain testimonial source, as we will see, and that might be avoided by employing justification defeaters.

If we connect the ideas above to argument mapping, we might make additional boxes (supporting boxes) to the boxes with judgments (or reasons) which are based on testimony. An example of an additional, supporting box would be testimony, which would be investigated separately in terms of justification (as a pro argument). What we need to determine in that investigation is when a belief (based on testimony) is justified in the intelligence context

---

8   See, for instance, Lackey 2003 and 2008, and Lackey and Sosa 2006. According to Burge (1993), it is a necessary truth, knowable a priori, that trust in testimonial sources is *prima facie* justified. Malmgren (2006) has argued against this.

in terms of testimony. That might involve a discussion on the track record of a source in terms of being right in the past. That might still cause problems, which has been discussed above. It would be better with justification defeaters.

Justification Defeaters/ Epistemic Defeaters[9]

There are two general types of defeaters. These can make a belief or knowledge claim lose its epistemic status or downgrade it: "propositional defeaters and mental state defeaters. Propositional defeaters are conditions external to the perspective of the cognizer that prevent an overall justified true belief from counting as knowledge" (Sudduth, 2024). As already mentioned, it is justification in general that is fruitful to use in the intelligence context— issues on truth and belief can be left aside. "Mental state defeaters are conditions internal to the perspective of the cognizer (such as experiences, beliefs, withholdings) that cancel, reduce, or even prevent justification" (Sudduth, 2024).

There is an idea in epistemology, that for a person's belief to be justified, there must be no "justification defeaters" for the belief. A justification defeater prevents a belief from being justified (Sudduth, 2024). Sudduth points out that this is what epistemologists call prima facie justification, and it means that when a belief is *prima facie* justified (justified on the assumption that there

are no defeaters) and there are in fact no defeaters, the result is *ultima facie* justification—or, justification all things epistemically considered. This is what we normally mean by "justified."

Prima facie justification plus no defeaters (required by no-defeaters clause) = *ultima facie* justification. That would not be quite relevant for the intelligence context, however.

Epistemologists sometimes distinguish between "rebutting defeaters" (somewhat similar to Toulmin's rebuttals (Toulmin, 2008, 93-95) and "undercutting defeaters." A rebutting defeater is illustrated in the following example from Sudduth (2024).

> Suppose you're in a pet shop where, in clear view, one of the store offerings catches your attention. "Tell me about that red dog," you say to the salesperson, who replies, "Oh, that's our new robotic pet. It's so realistic it fools everybody!" The testimony of the salesperson is a rebutting defeater of your prima facie justification for believing p ("that thing I'm looking at is a red-haired dog"); that is, the salesperson's testimony gives you good reason to think p is false. Thus, you lack ultima facie justification for believing p.

> An undercutting defeater is illustrated by altering the story slightly. Suppose the salesperson tells you there is a strong red light shining on the object

9   https://press.rebus.community/intro-to-phil-epistemology/chapter/epistemic-justification/

you're looking at. The salesperson's testimony is an undercutting defeater of your prima facie justification for believing p. This is because the salesperson gives you good reason to think the object would appear red even if it were not red. Even if the testimony does not give you good reason to think p is false, it does give you good reason to think the source of your belief (visual experience) is, in this situation, not good enough for ultima facie justification. (example from Sudduth 2024)

Sudduth points out that externalist theories (theories that base justification outside the epistemic subject (the person who knows something) "are in special need of a no-defeaters clause, without which they are susceptible to counterexamples." He gives the following example: "suppose your belief that p is produced by a reliable process type but you have good reason either to think that p is false or to think that the process producing your belief is unreliable. Without a no-defeaters clause, process reliabilism would implausibly imply that your belief is justified. Internalist evidentialist theories do not need a no-defeaters clause, because a person's total evidence at a time already weighs in any defeaters. Nevertheless, the notion of defeaters is sometimes employed by internalists as a useful tool for thinking about what a person's total evidence indicates" (Sudduth, 2024).

Justification Defeaters in the Argument Tree

As we have seen, reliability can be graded, like between A (completely reliable tried and trusted source that can be depended on with confidence) and E (unreliable source which has been used in the past and has proved unworthy of any confidence) and F (reliability cannot be judged—a source which has not been used in the past). The ambiguity of these criteria has been criticized, as well as how to deal with situations where sources have similar total evaluations like B3 and C2. Justification defeaters may not help with such specific evaluations, but they might help pinpointing weaknesses, point to the need for acquiring additional information and avoid pitfalls connected to sources, by being systematic.

Thomas Reid argues that we, by our nature, accept testimony unless we encounter special contrary reasons. Whether or not this is true, it seems there is good reason to making a point of looking for contrary reasons in the form of justification defeaters. This would be a good connection between philosophy/epistemology and intelligence analysis. To make a list of justification defeaters would be a way to make contrary reasons regarding testimony systematic.

One important factor that needs to be considered is the testimony-provider's (T = the testifier, S = the epistemic subject, that is, the person having knowledge (or trying to get knowledge)) ability to deceive—as well as their gullibility; neglecting the duty of monitoring signs of untrustworthiness. One way to find justification defeaters is to look for possible ulterior motives.

Arguments in favor of demands on T-based beliefs can be used as a list of things that can constitute justification defeaters. The list below is based on a list of arguments from Green (2024).

1. *T's ability to deceive.* The argument is that since testimony comes from a person, that is reason to be more demanding on testimonially-based beliefs compared to perceptually-based. (Faulkner, 2000; Lackey, 2006a, 176, 188 n. 44; Audi, 2006, 40): "(T) must in some sense select what to attend to."

2. *T's Irresponsibility and gullibility.* The argument, from Lackey (2006a) for instance, is that S (= the epistemic subject) needs positive reasons to believe T's testimony.

3. *S's ability not to trust T*—This is also something that needs to be considered.

4. *Operational dependence on other sources.* T as a source of beliefs requires other sources, like perception (see above—to each box based on T, we must look at the basic sources of knowledge (perception?) like memory, etc.

5. *Defeasibility of testimony*—based beliefs by other sources. Platinga (1993) and Audi (2006) suggest that testimony differs from sources like perception since T-based beliefs can be defeated by other sources

(Green, 2024).

6. *From a no-defeater condition to positive-reason-to-believe-condition*

7. *S's higher order beliefs about T.* When T tells S that p, one might demand that S has other beliefs concerning T or T's trustworthiness (Green, 2024).

The suggestion, to connect these ideas to the intelligence context, is the following. When the box in an argument tree has contents based on testimony, the idea is to make justificatory boxes for T as well as S, based on the list above, or a version of the list which might be relevant in this context—a context which makes epistemic justification even more difficult than it ordinarily is. The reason for this is the prevalence of phenomena like deception, misinformation and fabrication of intelligence, which is very much a part of the game. Whitesmith (2023, 838) also points out fabrication of intelligence for financial gain and the active attempts to keep information hidden by adversaries. This core of the intelligence context is an argument for looking extra at counter arguments and make this systematic. It is not quite sufficient with the suggestions from Bruce, Tucker, and Whitesmith, for instance. The most common risk of deception, misinformation and fabrication in the intelligence context is a support for making justification defeaters systematic in the argument tree, and the list above might be changed slightly to suit the context. This is a suggestion of a JD-list which should be connected to boxes with testimony.

JD-LIST FOR T-BOXES IN AN ARGUMENT TREE

- **1: T's (T = the testifier) motive and ability to deceive.** We should add "motive" here, for reasons mentioned above. May T have a motive to deceive? What about the possibility to deceive due to negligence? See the following point in the list. Audi (2006, 40) "(T) must in some sense select what to attend to." (Green, 2024)

- **2: T's irresponsibility and gullibility**: Connected to Bruce's habit of thought. Lackey (2006a) argues that S needs positive reasons to believe T's testimony. Fricker (1994) argues that S has a duty to monitor T for signs of untrustworthiness (Green, 2024). Also, can T deceive due to negligence/irresponsibility? Has T been negligent in the past?

- **3a: S's (S = the epistemic subject) ability not to trust T.** Audi (2006, 40) argues that "(S) commonly can withhold belief, if not at will then indirectly, by taking on a high-

ly cautionary frame of mind." A highly cautionary mind seems a prerequisite for dealing with the entire list (Green, 2024). Move up the point of higher order beliefs as a supporting box.

- **3b: (S's higher order beliefs about T).** When T tells S that p, one might demand that S have other beliefs concerning T or T's trustworthiness.

The suggestion is to remove *Operational dependence on other sources*, as well as *defeasibility of testimony*, from the list, since "sources" has to do with things like perception in philosophy. Also remove *From a no-defeater condition to positive-reason-to-believe-condition*. Counter arguments from other sources, are separate boxes in the tree. So the question is how to come up with counter arguments such as C1P1—how to be able to place such a statement in a tree, since the fact that Curveball was exposed as a fabricator was not known at the time. The suggestion is to use a JD-list for box P1 and boxes connected to this.

Main claim: **Iraq possesses CV (chemical weapons)**

P1 (first pro argument for main claim): (NIE): Iraq has an active CV program involving the production of mustard, sarin, cyclosarin and VX and has stockpiled 100-500 metric tons

P1P1 (first pro argument for P1): Images of Samarra type tanker trucks. (Bruce: simplified, fully elaborated in WHO commission report, 122-124).

P1P1P1 (first pro argument for P1P1): Imagery analysts says so (authority – a combination of empirical observations and expert judgment).

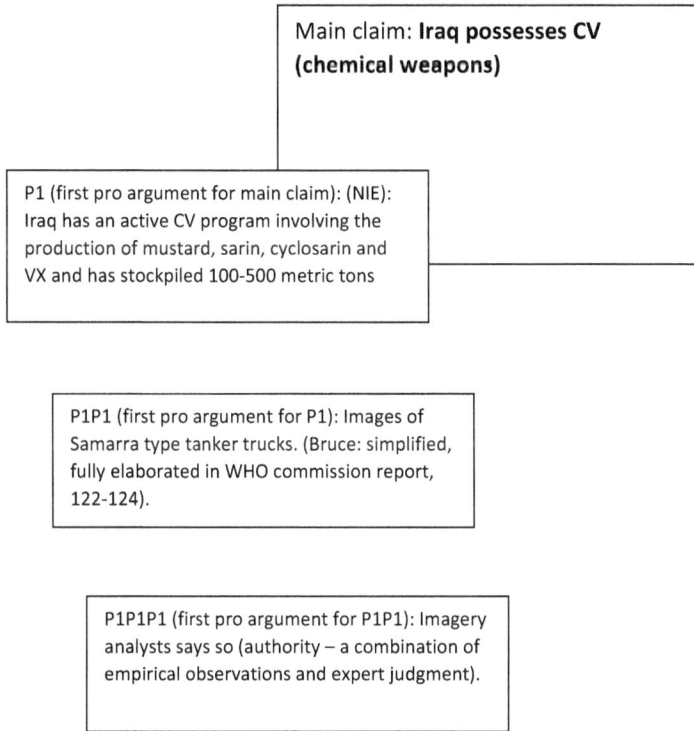

ARGUMENT TREE 2

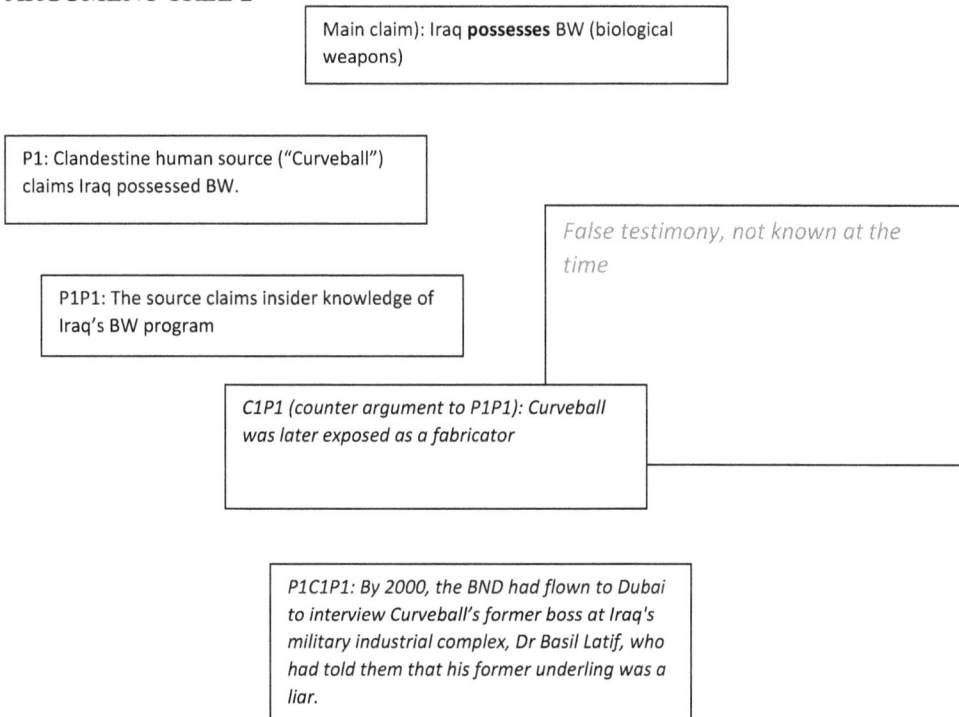

Main claim): Iraq **possesses** BW (biological weapons)

P1: Clandestine human source ("Curveball") claims Iraq possessed BW.

False testimony, not known at the time

P1P1: The source claims insider knowledge of Iraq's BW program

C1P1 (counter argument to P1P1): Curveball was later exposed as a fabricator

P1C1P1: By 2000, the BND had flown to Dubai to interview Curveball's former boss at Iraq's military industrial complex, Dr Basil Latif, who had told them that his former underling was a liar.

Boxes P1 and C1P1. Would that be captured by a JD-list? Curveball claimed to have inside knowledge about Iran's BW program (P1P1 box).

The idea is that by using the JD-list, the risk of a thing like C1P1 would be less easily overlooked. That is, C1P1 would—at the time—consist of the JD-list.

JD-LIST FOR BOXES CONNECTED TO P1-BOX

- **1: T's (T = the testifier) motive and ability to deceive**. Look closely into Curveball's motive and ability to deceive. Think of possible motives for Curveball to deceive. Write this down in a C1P1-box, point 1.

- **2: T's irresponsibility and gullibility**: Look closely into Curveball's signs of untrustworthiness, perhaps in other areas. Har Curveball been negligent in the past? S (the epistemic subject) has a duty to monitor Curveball for such signs. S also needs positive reasons to believe Curveball's testimony, make those explicit.

- **3a: S's (S = the epistemic subject) ability not to trust T.** Audi (2006, 40) argues that "(S) commonly can withhold belief, if not at will then indirectly, by taking on a highly cautionary frame of mind." A highly cautionary mind seems a prerequisite for dealing with the entire list, based on Audi's reasoning on S keeping a highly cautionary frame of mind (Audi, 2006, 40; Green, 2024). This is a point of

higher order beliefs about oneself, that is S's beliefs about S's owns belief and ability to detect issues connected to 1 and 2 on the list.

- **3b: (S's higher order beliefs about T).** When T tells S that p, one might demand that S have other beliefs concerning T or T's trustworthiness as in point 3a.

The rest of the list may constitute boxes with counter arguments. But this list, with points 1-3, should help detect weaknesses in the argument connected to testimony by making it explicit and systematic.

Conclusion

Since there are different contents in the boxes of an argument tree, there are different criteria for when a belief is justified. Testimonial evidence is central for intelligence, and the most difficult when it comes to justification not only because it is prone to error, but also because of the special features of the intelligence context in terms of active deception and fabrication, for instance.

In order to deal with the boxes in the argument tree which are based on testimony, it is necessary to deal with this special context in regard to justification and determining whether a belief is justified. The suggestion is to have an extended argument tree—to lay out and thoroughly explore so called justification defeaters connected to testimony. To do this in a systematic manner, one can add boxes to the boxes based on testimony. These boxes should contain a list of justificatory defeaters.

The question this paper aimed to answer was: How can the philosophical notion of justification ("justified belief") be understood and used in the intelligence context connected to testimony and evaluating sources? The suggestion is: as a belief where the justification defeaters have been taken into account explicitly.

Linda Johansson holds a PhD in Philosophy from the Royal Institute of Technology in Stockholm, Sweden. She is a teacher and researcher at the Swedish Defence University. Her primary area of research includes intelligence analysis and autonomous systems, with a focus on philosophical aspects. She welcomes opportunities for continued research and collaboration.

References

Audi, R. (2002). The sources of knowledge. In P. Moser (Ed.), The Oxford handbook of epistemology (pp. 71–94). New York: Oxford University Press.

Audi, Robert. (2006). "Testimony, Credulity, and Veracity," in Lackey and Sosa 2006.

Barbieri, D. (2013). Bayesian intelligence analysis. Intelligence In the Knowledge Society, Proceedings of the XIXth International Conference. Bucharest.

BonJour, L. (2010). Epistemology: Classic problems and contemporary responses (2nd ed.). Lanham, MD: Rowman & Littlefield.

Bruce, J. B. (2008). Making analysis more reliable: Why epistemology matters to intelligence. In George, R. Z. and Bruce, J. B. (eds), Analyzing Intelligence: Origins, Obstacles, and Innovations, Washington D.C.: Georgetown University Press, pp. 171-90.

Clark, Robert M. (2020). *Intelligence Analysis. A Target Centric Approach*. CQ Press, California.

Faulkner, Paul. (2000). "The Social Character of Testimonial Knowledge." *Journal of Philosophy* 97: 581-601.

Fricker, Elizabeth. (1994). Against Gullibility, in *Knowing from Words: Western and Indian Philosophical Analysis of Understanding and Testimony*, Bimal Krishna Matilal and Arindam Chakrabarti (eds.), Dordrecht: Springer Netherlands, 125–161.

Green, Christopher R. (2006). *The Epistemic Parity of Testimony, Memory, and Perception*. Ph.D. dissertation, University of Notre Dame.

Green, Christopher R. (2007). "Suing One's Sense Faculties for Fraud: 'Justifiable Reliance' in the Law as a Clue to Epistemic Justification." *Philosophical Papers* 36: 49-90.

Green, Christopher R. (2024). "Epistemology of Testimony." *The Internet Encyclopedia of Philosophy*, ISSN 2161-0002, https://iep.utm.edu/, 2024-09-30.

Hendrickson, Noel. (2018). *Reasoning for Intelligence Analysts*. Rowman & Littlefield.

Heuer, Richards J. (1978). ed., Quantitative Approaches to Political Intelligence: The CIA experience *Westview Press*, Boulder Colorado, 1978: 4-10.

Heuer, Richards J. (1999). Psychology of Intelligence Analysis (Washington D.C., Center for the Study of Intelligence, Central Intelligence Agency, 1999), pp. 77-88.

Heuer, Richards J. & Pherson, Randolph H. (2010). *Structured Analytic Techniques for Intelligence Analysts*. Thousand Oaks, CA: CQ Press.

Irwin, D., & Mandel, D. R. (2019). Improving information evaluation for intelligence production. Intelligence and National Security, 34(4), 503–525.

Johansson, L. (2024). Argument mapping in intelligence analysis. Forthcoming in *Security and Intelligence*, vol 9, no 2.

Lackey, Jennifer. (2006a). "It Takes Two to Tango: Beyond Reductionism and Non-Reductionism in the Epistemology of Testimony," in Lackey and Sosa 2006.

Plantiga, A. (1993). Warrant and proper function. Oxford: Oxford University Press

Steup, Matthias and Ram Neta. (2020). Epistemology. *The Stanford Encyclopedia of Philosophy* (Fall 2020 Edition), Edward N. Zalta (ed.), URL = <https://plato.stanford.edu/archives/fall2020/entries/epistemology/>.

Sudduth, M. (2024). "Defeaters in Epistemology" *The Internet Encyclopedia of Philosophy*, ISSN 2161-0002, https://iep.utm.edu/, 2024-09-30.

Thagard, P. (2005). Testimony, credibility, and explanatory coherence. Erkenntnis, 63, 295–316.

Tucker, Avezier. (2023). From unreliable sources: Bayesian critique and normative modelling of HUMINT inferences, *Journal of Policing, Intelligence and Counter Terrorism*, DOI: 10.1080/18335330.2023.2187704

Tecuci, Gheorghe, Schum, David A., Marcu, Dorin, and Boicu, Mihai. (2016). *Intelligence Analysis as Discovery of Evidence, Hypothese and Arguments*, New York: Cambridge University Press, 2016.

Toulmin, Stephen. (2008). *The Use of Argument*, New York: Cambridge University Press.

Van Gelder, Timothy. (2007). The Rationale for Rationale. *Law, Probability and Risk* 6 (2007): 23-42.

Wheaton, Kristian J., Lee, Jennifer, and Deshmukh, Hemangini. (2010). Teaching Bayesian Statistics to Intelligence analysts: Lessons learned. *Journal of Strategic Security* 2, no. 1 (2010): 39-58.

Whitesmith, Martha. (2022a). *Cognitive Bias in Intelligence Analysis – Testing the Analysis of Competing Hypotheses Method*. Edinburgh: Edinburg University Press.

Whitesmith, Martha. (2022b). Justified true belief theory for intelligence analysis. *Intelligence and National Security*, 37:6, 835-849.

Security and Intelligence • Volume 10, Number 1 • Spring 2025

Kinetics to Keyboards: The Future of the Special Operations Force (SOF) Operator Post Global War on Terror (GWOT)

AJ Rutherford[1]

ABSTRACT

The discussion underscores the critical importance of interdisciplinary training, including cybersecurity, digital forensics, and network penetration, as well as the need for collaboration with government agencies, private industry, and academia. Ethical and legal challenges surrounding offensive cyber operations, such as compliance with international law and sovereignty, are also explored. Furthermore, the role of emerging technologies such as artificial intelligence (AI), machine learning (ML), and quantum computing in enhancing SOF operational agility is addressed. The findings suggest that the integration of cyber warfare into SOF missions acts as a force multiplier, enabling precision, scalability, and strategic flexibility in increasingly contested environments. However, the reliance on digital infrastructure introduces vulnerabilities, highlighting the need for robust cybersecurity measures and continuous adaptation to technological advancements. This article concludes by offering policy recommendations for advancing SOF training, resource allocation, and joint cyber operations, while identifying areas for future research, including the long-term impact of cyber warfare on SOF and other military units.

Keywords: Special Operations Forces; Cyber effects; Future War; Artificial Intelligence; Machine Learning

De la cinética al teclado: El futuro del operador de las Fuerzas de Operaciones Especiales (FOE) tras la Guerra Global contra el Terror (GWOT)

RESUMEN

El debate subraya la importancia crucial de la formación interdisciplinaria, que incluye ciberseguridad, análisis forense digital y penetración de redes, así como la necesidad de colaboración con

---
1   ajrutherford0372@gmail.com

doi: 10.18278/si.10.1.4

organismos gubernamentales, la industria privada y el mundo aca-
démico. También se exploran los desafíos éticos y legales que ro-
dean las operaciones cibernéticas ofensivas, como el cumplimiento
del derecho internacional y la soberanía. Además, se aborda el papel
de las tecnologías emergentes, como la inteligencia artificial (IA), el
aprendizaje automático (AA) y la computación cuántica, en la me-
jora de la agilidad operativa de las FOE. Los hallazgos sugieren que
la integración de la ciberguerra en las misiones de las FOE actúa
como un multiplicador de fuerza, permitiendo precisión, escala-
bilidad y flexibilidad estratégica en entornos cada vez más contro-
vertidos. Sin embargo, la dependencia de la infraestructura digital
introduce vulnerabilidades, lo que pone de relieve la necesidad de
medidas robustas de ciberseguridad y una adaptación continua a los
avances tecnológicos. Este artículo concluye ofreciendo recomendaciones de políticas para impulsar el entrenamiento de las Fuerzas de
Operaciones Especiales (FOE), la asignación de recursos y las operaciones cibernéticas conjuntas, a la vez que identifica áreas de investigación futura, incluyendo el impacto a largo plazo de la ciberguerra
en las FOE y otras unidades militares.

Palabras clave: Fuerzas de Operaciones Especiales; Efectos ci-
bernéticos; Guerra del futuro; Inteligencia artificial; Aprendizaje
automático

从动力学到键盘：全球反恐战争(GWOT)后
特种作战部队(SOF)操作者的未来

摘要

本篇讨论文强调了跨学科培训（包括网络安全、数字取证和
网络渗透）的重要性，以及与政府机构、私营企业和学术界
合作的必要性。此外，本文还探讨了围绕进攻性网络作战的
伦理和法律挑战，例如遵守国际法和维护主权。此外，本文
还探讨了人工智能(AI)、机器学习(ML)和量子计算等新兴技
术在提升特种作战部队操作敏捷性方面的作用。研究结果表
明，"将网络战融入特种作战部队任务"充当了力量倍增器
的作用，在竞争日益激烈的环境中实现精准度、可扩展性和
战略灵活性。然而，对数字基础设施的依赖也带来了漏洞，
强调需要采取强有力的网络安全措施和持续适应技术进步。
本文最后提出了推进特种作战部队训练、资源配置和联合网
络作战的政策建议，并确定了未来的研究领域，包括网络战
对特种作战部队和其他军事单位的长期影响。

关键词：特种作战部队，网络效应，未来战争，人工智能，机器学习

Introduction

A. Background on Special Operations Forces (SOF)

Special Operations Forces (SOF) have become a critical element in modern military conflicts, performing specialized, high-impact operations that conventional military forces are often ill-equipped to undertake. These forces excel in complex environments where flexibility and precision are paramount. Key functions of SOF include counterterrorism, unconventional warfare, direct action, and special reconnaissance, which allow them to operate in hostile territories and unstable political environments with agility. SOF units are particularly effective in irregular warfare, where they often engage in small, agile teams capable of infiltrating denied areas, collecting intelligence, and executing precise strikes [1]. This suggests that increasing cyber capabilities within SOF would enhance their effectiveness in irregular warfare. By integrating advanced cyber tools and skills, SOF could improve their ability to infiltrate denied areas digitally, collect intelligence through cyber means, and execute precise cyber strikes. This would further increase their agility, adaptability, and operational effectiveness in dynamic and unpredictable environments. Additionally, enhanced cyber capabilities would enable SOF to disrupt adversaries' operations, gather critical data, and provide real-time tactical support.

SOF's strategic role also extends beyond tactical operations, as they frequently collaborate with indigenous forces to strengthen local governance, bolster security efforts, and stabilize regions affected by insurgencies. The emphasis that SOF units are not only involved in military engagements but also in capacity-building efforts with partner nations, ensuring the sustainable security of volatile regions [2]. The implication of SOF's capacity-building efforts is enhanced through cyber skills by enabling SOF units to engage in cyber operations that bolster local and regional security frameworks. Cyber skills allow SOF to support partner nations in defending against cyber threats, securing critical infrastructure, and enhancing communication systems. By providing training, tools, and expertise in cybersecurity, SOF can help partner forces identify, respond to, and mitigate digital threats, thus strengthening their resilience in an increasingly interconnected and cyber-dependent world. This aligns with SOF's broader goal of sustainable security, ensuring that partners are capable of defending against both conventional and cyber threats. Through these strategic partnerships, SOF can engage in operations that extend beyond military actions, contributing to broader geopolitical stability and fostering long-term peace in areas plagued by insurgency or instability. The integration of Special Operations Forces (SOF) with national-level cyber architectures, such as the Cyber National Mission Forces (CNMF), represents a transformative evolution in modern warfare, enabling a seamless fusion of cyber and kinetic operations.

As a critical component of U.S. Cyber Command (USCYBERCOM), CNMF comprises specialized mission teams, including National Mission Teams (NMTs) tasked with countering nation-state cyber threats, Cyber Protection Teams (CPTs) responsible for securing Department of Defense (DoD) networks, and Combat Mission Teams (CMTs) that conduct offensive cyber operations in support of joint force objectives [3]. This structure provides SOF with a force-multiplying capability, extending their operational reach through cyber-enabled reconnaissance, electronic warfare, and adversary network disruption. CNMF's advanced cyber capabilities allow SOF to conduct multi-domain operations by leveraging digital reconnaissance for enhanced situational awareness, deploying offensive cyber tools to disable enemy command-and-control systems, and securing battlefield networks against intrusion and electronic warfare threats [4]. Additionally, this integration aligns with the Joint All-Domain Command and Control (JADC2) framework, ensuring real-time intelligence fusion and rapid decision-making for SOF missions in contested environments [5].

By embedding cyber operations within SOF mission sets, the United States strengthens its capacity to execute asymmetric warfare strategies, where cyber and information operations complement traditional direct-action missions. This paradigm shift necessitates revised doctrine, expanded interagency collaboration, and advanced training initiatives to equip SOF personnel with the technical expertise required for cyber-enabled special operations. As adversaries increasingly exploit digital battlefields, the CNMF-SOF synergy underscores the imperative for an integrated approach to modern warfare, ensuring that SOF remains agile, resilient, and capable of executing operations across both physical and virtual domains.

B. Historical Evolution from Conventional to Asymmetric and Unconventional Warfare

The evolution of warfare strategies from conventional to asymmetric and unconventional tactics reflects the shifting dynamics of global conflict over the past century. Conventional warfare, characterized by large-scale engagements between regular armies, is increasingly less effective against insurgent groups and non-state actors who leverage asymmetrical tactics such as guerrilla warfare, sabotage, and psychological operations [6] [7]. Asymmetric warfare has become a common response to the technological and tactical advantages of conventional forces, with smaller, less-equipped groups utilizing unconventional strategies to disrupt more powerful opponents. This suggests that in the face of technologically superior and better-equipped conventional forces, smaller or less-equipped groups are increasingly adopting asymmetric warfare tactics. These tactics allow these groups to exploit unconventional strategies such as cyberattacks, guerrilla warfare, or misinformation campaigns to disrupt, undermine, and challenge the larger, more powerful opponents in a way that negates the latter's advantages.

The limitations of conventional strategies were starkly illustrated during the Vietnam War, where a determined insurgency successfully leveraged local support, terrain, and guerrilla tactics to counter a technologically superior military force [8]. This same observation can be made with the Global War on Terror in both Iraq and Afghanistan. This shift from territorial conquest and large-scale battles to strategies prioritizing mobility, intelligence, and psychological operations signals a critical change in the way modern conflicts are fought. The focus has increasingly moved toward smaller, more adaptable forces capable of employing irregular tactics that challenge traditional military paradigms [9].

C. The Rise of Cyber Warfare

Cyber warfare, the use of digital technologies as tools of conflict, has emerged as a pivotal dimension in modern military strategies. Cyber warfare is defined by activities such as hacking, denial-of-service (DoS) attacks, cyber espionage, and the deployment of malware, cyber warfare enables states and non-state actors to disrupt critical infrastructure, steal sensitive information, and gain strategic

advantages over adversaries [10] [11]. Unlike traditional warfare, cyber operations can be executed without physical force, complicating attribution and accountability and raising significant challenges in the areas of international law and policy [12]. This argues that cyber operations differ from traditional warfare in that they do not rely on physical force, making it difficult to attribute attacks to specific actors and hold them accountable. This lack of physical evidence complicates the application of international law and policy, creating significant challenges in defining and responding to cyber conflicts.

The non-kinetic nature of cyber warfare blurs the lines between peacetime and conflict, as digital attacks are often conducted in the gray zone of international relations. The growing reliance on digital infrastructure—critical to national security and economic systems—has only increased the prominence of cyber warfare, underscoring the need for robust defense mechanisms and international cooperation in the digital space.

D. Impact of Digital Transformation on Global Conflict and Military Operations

The digital transformation has reshaped the landscape of global conflict and military operations, introducing new technologies that have altered both the nature of warfare and the strategies employed by military forces. Advancements in artificial intelligence (AI), machine learning (ML), cyber capabilities, and autonomous systems have not only improved military effectiveness but also introduced new vulnerabilities that adversaries can exploit. Current trends emphasize on how cyber capabilities have become essential tools in modern power projection, enabling state and non-state actors to influence geopolitical dynamics through digital attacks and disinformation campaigns [11]. An example of this is Russia's interference in the 2016 United States presidential election. Through cyber capabilities, Russian state-sponsored hackers conducted digital attacks to breach Democratic National Committee (DNC) systems and spread disinformation via social media. This cyber operation aimed to influence public opinion, destabilize the political landscape, and exert geopolitical influence without physical military force. It highlights how cyber capabilities are now critical tools for influencing geopolitics, with state and non-state actors leveraging them to advance their strategic goals.

Cyberspace is increasingly critical to military operations, where the use of cyber tools can disrupt communication networks, disable infrastructure, and degrade the effectiveness of traditional military strategies [10]. However, these technological advancements also pose a significant risk, making cybersecurity an urgent priority to safeguard military assets from increasingly sophisticated digital threats. The rise of cyber warfare has not only introduced a new mode of attack but has also forced a reconsideration of existing military strategies, highlighting the need for new defense frameworks in this digital age.

E. Purpose and Scope of the Article

This article explores how cyber warfare is reshaping the roles and operational strategies of SOF. As military conflicts increasingly involve both traditional kinetic operations and digital warfare, SOF must adapt by incorporating cyber capabilities into their mission set. SOF are now tasked with operations such as cyber intelligence gathering, disrupting adversary networks, and defending against influence campaigns in hybrid conflict scenarios. The article discusses the evolution of SOF tactics in response to cyber threats, emphasizing the importance of cyber integration into their missions and highlighting the need for specialized training and collaboration with cyber professionals.

Additionally, the article examines the future of SOF in a cyber-dominated battlefield, exploring the evolving technological capabilities that SOF will require to remain effective. These include advanced proficiency in cyber warfare, electronic warfare, and cognitive dominance, with particular attention to the integration of AI and emerging technologies that could transform the tactical landscape.

Literature Review

The Role of Cybersecurity in the Evolution of Modern Warfare

The integration of cybersecurity into military operations has become a critical facet of modern warfare. With the increased reliance on digital infrastructure, nations are now compelled to address the evolving threats posed by cyber actors. This literature review aims to explore the changing nature of warfare, particularly focusing on the increasing role of cybersecurity and cyber operations in modern conflicts. It draws upon studies from diverse disciplines to understand the impact of cyber warfare on traditional military strategies, the ethical implications, and the necessity for robust cybersecurity measures to support both conventional and non-conventional military operations.

Cyber Warfare as an Emerging Domain in Modern Conflicts

Cyber warfare has rapidly emerged as a significant domain of military conflict, transforming how nations engage in and approach warfare. The digital age has led to a paradigm shift where cyber operations are integrated into traditional military strategies [10]. The rise of cyber operations has rendered the conventional boundaries of war less defined, as cyberattacks can occur alongside kinetic operations, potentially without physical engagement. For example, cyber tools such as malware, Denial of Service (DoS) attacks, phishing, and social media manipulation can achieve strategic objectives covertly, cost-effectively, and without the destruction associated with conventional warfare [13]. Reinforcing this point, cyber capabilities empower states to leverage "soft power," engaging in tactics such as disinformation campaigns, economic sabotage, and targeted disruption of critical infrastructure, all of which destabilize adversaries without physical conflict [11].

The application of cyber capabilities, such as data breaches, ransomware, and critical infrastructure attacks, demonstrates the growing vulnerability of modern societies to non-kinetic warfare. Illustrating this is the vulnerability of the Ukrainian power grid to Russian-backed cyberattacks, where cyber tools effectively disabled vital services, creating national destabilization without the need for conventional military engagement [14]. These incidents highlight the multifaceted nature of modern warfare, where cyberattacks are often strategically planned to achieve objectives that would otherwise require physical military action. This suggests that modern warfare increasingly relies on cyberattacks as a primary tool for achieving strategic objectives. It implies that cyberattacks can be used to accomplish goals—such as disabling infrastructure, stealing sensitive information, or influencing public opinion—that would traditionally require physical military action. Cyber capabilities provide a cost-effective, covert, and often less risky alternative to conventional warfare, enabling actors to achieve their aims without direct confrontation or destruction.

Cybersecurity: A Critical Component of Military Operations

As warfare becomes increasingly reliant on cyber capabilities, the role of cybersecurity in military operations cannot be overstated. The shift from traditional warfare to hybrid and asymmetric threats has necessitated the integration of cybersecurity into military doctrine. The future of military strategy is deeply intertwined with cyber warfare, and effective cybersecurity frameworks must be adopted to support both offensive and defensive operations [15]. The ability to secure communications, protect sensitive data, and safeguard critical infrastructure is essential for operational success, especially as adversaries continue to exploit technological vulnerabilities.

Moreover, the ethical and legal complexities surrounding offensive cyber operations have sparked considerable debate. Issues of attribution, sovereignty, and the application of international law are paramount when discussing cyber warfare. The difficulty in establishing norms and rules of engagement in cyberspace, where actions can be anonymous, and responses often lack clear legal frameworks. The need for governments and military organizations to develop comprehensive policies and guidelines for cyber operations is essential to maintaining international peace and security while mitigating the potential for unintended escalations in conflicts.

Case Studies: Ukraine-Russia Conflict and Operation Orchard

The evolving role of cyber warfare is evident in several key case studies, most notably the Ukraine-Russia conflict and Israel's Operation Orchard. The 2015 cyberattack on the Ukrainian power grid remains one of the most prominent examples of how cyber operations can disrupt an adversary's critical infrastructure, causing economic and social turmoil. Russian-backed hackers

targeted the Ukrainian electrical grid, causing a massive blackout and demonstrating the capability of cyber weapons to inflict substantial damage without the use of conventional military force [14]. This case exemplifies the growing significance of cybersecurity as an integral component of national defense.

Similarly, Israel's Operation Orchard in 2007, which allegedly involved cyberattacks on Syrian nuclear facilities, showcased the strategic advantage of cyber operations. Israel reportedly used cyber tools to disable Syrian radar systems before launching a military strike on the facility, preventing Syria from developing weapons of mass destruction. The shows that the integration of cyber capabilities into military operations offers a scalable, low-cost means to achieve strategic goals while maintaining plausible deniability [15]. These examples highlight how cyber operations can complement traditional military tactics, enhancing the scope of strategic objectives in modern conflicts.

Discussion

Integration of Cyber Capabilities into Special Operations Forces

SOF have increasingly embraced cyber capabilities to enhance their operational agility and effectiveness. The need for SOF to operate in both the physical and digital domains has led to the needed development of cyber-SOF operators who possess both traditional combat and cyber warfare expertise. SOF's involvement in cyber operations includes targeting adversary networks, securing digital communication channels, and leveraging data analytics to support mission planning [16]. This dual proficiency enables SOF units to conduct operations in a multidimensional battlefield, incorporating both kinetic and cyber actions to achieve operational success.

The integration of cyber operations into SOF missions requires overcoming significant technical and organizational challenges. Stressing the importance of interoperability between SOF and cyber units, the need for advanced training programs and collaborative operational frameworks [17] between the two suggests it is necessary to introduce cyber technical capabilities that can be achieved through specific cyber industry certifications (Certified Ethical Hacker – CEH, Offensive Security Certified Professional – OSCP, Certified Information System Security Professional – CISSP etc.) for the operator. This integration ensures that SOF units can seamlessly conduct operations across both cyber and physical domains, increasing mission flexibility and responsiveness to emerging threats.

Emerging Technologies and the Future of Cyber Warfare

Looking to the future, emerging technologies such as artificial intelligence (AI), machine learning (ML), and quantum computing are poised to play a pivotal role in enhancing cyber warfare capabilities. AI and ML can be employed to analyze large datasets and predict adversarial actions, providing military forces with a strategic edge in both defense and offense. Highlighting the potential of AI in automating threat detection and enhancing cyber defense systems, AI can reduce the risk of cyberattacks in critical infrastructure [11]. AI can also be used to effectively detect anomalies in intelligence datasets providing greater visibility of the battlespace at a more rapid rate providing a more current view of the battlefield allowing operators to make more informed decisions quicker. Additionally, quantum computing promises to revolutionize encryption techniques, offering more robust methods to protect sensitive information from advanced cyber threats.

However, these technologies also present new vulnerabilities. The widespread adoption of AI and quantum computing may lead to new forms of cyberattacks that exploit weaknesses in current security measures [17]. An example that illustrates this statement about the potential risks of AI and quantum computing in cyberattacks is the threat posed by quantum computers to current encryption methods, particularly public-key cryptography systems like RSA and ECC (Elliptic Curve Cryptography) which directly impact data protection, secure communications, and operational security (OPSEC).

Quantum computers, leveraging principles of quantum mechanics, could theoretically break existing encryption protocols much faster than classical computers. Shor's Algorithm, a quantum algorithm, can efficiently factor large numbers, which is the basis of RSA encryption, and solve discrete logarithms, which underpins ECC. This could render currently used encryption methods obsolete and vulnerable to cyberattacks by adversaries with access to quantum computing capabilities.

For example, the U.S. National Security Agency (NSA) has acknowledged the potential risks posed by quantum computing to encryption standards and has begun exploring post-quantum cryptography to develop encryption methods that are resistant to quantum computing threats.

This potential threat illustrates how quantum computing, once fully realized, could exploit weaknesses in current cryptographic systems and drastically change the landscape of cyberattacks. Attackers with access to powerful quantum computers could decrypt sensitive data, such as classified information, financial transactions, and private communications, rendering existing cybersecurity measures ineffective and forcing a major overhaul of encryption standards and security infrastructure.

Consequently, it is essential for military organizations to continuously adapt to technological advancements and invest in cutting-edge cybersecurity tools to stay ahead of adversarial capabilities.

The integration of cyber warfare into modern military strategies has fundamentally transformed the way conflicts are waged. As cyber tools provide strategic advantages without physical destruction, they have reshaped traditional concepts of warfare. The case studies of Ukraine-Russia and Israel's Operation Orchard illustrate the growing significance of cyber operations in achieving military objectives. To effectively integrate cyber capabilities into military doctrine, robust cybersecurity measures, interdisciplinary training, and international cooperation are essential. As emerging technologies continue to advance, the evolution of cyber warfare will require continuous adaptation to maintain strategic superiority in an increasingly contested digital landscape.

The Role of the SOF Operator in the Cyber-Dominated Battlefield: Expanding Capabilities and Integrating Domains

As modern warfare increasingly intertwines digital technologies with traditional military operations, the role of SOF must evolve. Historically, SOF units have excelled in executing high-stakes, high-risk missions requiring precision and flexibility. However, the evolving nature of cyber warfare continues to present complex challenges and opportunities for SOF operators. In response to the growing prominence of cyber threats, SOF must develop proficiency not only in conventional combat skills but also in specialized cyber operations. This integration of cyber capabilities is not merely an adjunct to traditional tactics but a fundamental shift that enhances SOF's effectiveness across multiple domains. The argument exists that SOF units must master both conventional and cyber warfare to stay competitive in the rapidly evolving battlefield [17]. Such dual proficiency allows SOF to operate seamlessly across the kinetic and cyber domains, improving operational flexibility and mission success as well as maintaining relevancy in the current battlespace.

Expanding the SOF Skillset for the Digital Age

The need for proficiency in cyber operations alongside traditional SOF skills has become imperative as adversaries increasingly exploit digital vulnerabilities. Comparisons of traditional SOF skills (specifically reconnaissance which supports all other SOF mission sets) translated to the more modern cyber skills are illustrated in the figure below:

Figure 1: Traditional SOF to Cyber SOF Skills

TRADITIONAL SKILL Physical Reconnaissance SOF	CYBER SKILL Cyber Reconnaissance Ethical - Hacking/ Pentesting	DESCRIPTION AND COMPARISON
Observation and Surveillance	Passive Reconnaissance (e.g., WHOIS, Shodan)	Just as SOF personnel would observe an area for intel without directly interacting with it, passive cyber reconnaissance gathers information without alerting the target, using open-source intelligence (OSINT) tools like WHOIS to map a target's digital footprint.
Maps and Satellite Imagery Analysis	Network Mapping (e.g., Nmap, Censys)	SOF uses maps and satellite imagery to understand terrain and infrastructure. Similarly, ethical hackers use tools like Nmap and Censys to visualize and analyze network structures, IP addresses, and system configurations of a target environment.

(Cont'd.)

TRADITIONAL SKILL Physical Reconnaissance SOF	CYBER SKILL Cyber Reconnaissance Ethical - Hacking/ Pentesting	DESCRIPTION AND COMPARISON
Listening to Enemy Radio Traffic (SIGINT)	Traffic Analysis (e.g., Wireshark)	SIGINT gathers communications without interacting directly. In cyber, traffic analysis with Wireshark can passively monitor network traffic and capture packets, providing insight into potential vulnerabilities.
Infiltration to Access Critical Areas	Active Scanning (e.g., Nessus, Masscan)	SOF teams may physically infiltrate to observe and map critical infrastructure. Active scans in cyber recon, like those from Nessus or Masscan, "probe" network systems to reveal weaknesses, much like direct infiltration provides detailed intel on a physical structure's vulnerabilities.
Stakeouts (Observation Posts or OPs – Hide Sites) for Target Behavior Monitoring	Tracking Devices and Services (e.g., Google Dorking, theHarvester)	Ops allow SOF to observe patterns in target behavior. Google Dorking and theHarvester help track an organization's online presence, such as email addresses, domains, and publicly available data, identifying patterns and potential entry points.
Identification of Guard and Patrol Patterns	Identification of Security Mechanisms (e.g., Firewalls, IDS)	SOF assesses guard routines to time movements. Ethical hackers assess security mechanisms, such as firewalls and intrusion detection systems, to understand and possibly bypass defenses at the optimal time.

(Cont'd.)

TRADITIONAL SKILL Physical Reconnaissance SOF	CYBER SKILL Cyber Reconnaissance Ethical - Hacking/ Pentesting	DESCRIPTION AND COMPARISON
Photography and Document Collection	Data Extraction (e.g., Maltego)	SOF may photograph or document target information, while ethical hackers use tools like Maltego to visualize and connect entities (e.g., emails, domains, people) from extracted data, building a detailed understanding of an organization's digital infrastructure.
Using Cover and Concealment for Covert Movement	Anonymized Recon (e.g., Tor, Proxychains)	SOF uses terrain and cover to avoid detection during physical recon. Cyber operators use Tor and proxychains to anonymize their activities, masking their IP addresses to evade detection by the target network.
Interrogation of Locals for Intel	Social Engineering (e.g., Phishing)	SOF may gather intel by speaking with local contacts. Similarly, ethical hackers might use social engineering tactics, such as phishing, to gather intel from users of the target system, although this is a separate skillset from pure technical recon.
Hand-Drawing Facility Layouts	Network Diagrams (e.g., manually or with tools)	SOF may create physical sketches of a building's layout. In cyber recon, the analyst may manually draw network diagrams or use mapping tools to structure data gathered, outlining network pathways and device locations virtually.

(Cont'd.)

TRADITIONAL SKILL Physical Reconnaissance SOF	CYBER SKILL Cyber Reconnaissance Ethical - Hacking/ Pentesting	DESCRIPTION AND COMPARISON
Camouflage and Concealment	Obfuscation Techniques (e.g., slow scanning)	SOF may use camouflage to blend in. Ethical hackers use obfuscation techniques, like conducting slow or random scans, to reduce the risk of triggering alerts in active systems, making recon appear as background traffic.

As SOF units engage in complex, multi-domain operations (MDO), their ability to disrupt enemy communications, protect digital infrastructure, and execute cyber-attacks and other cyber focused capabilities will be vital. The integration of cyber operations into SOF missions is necessary for maintaining strategic, operational, and tactical advantage. The ability to carry out cyber operations—including disabling critical enemy infrastructure and digital targeting—complements traditional SOF tactics like direct action and reconnaissance, creating a more flexible and adaptive force [19].

The importance of establishing joint operational frameworks where cyber expertise and kinetic capabilities converge is essential for integrated success of both the kinetic and cyber domains to build a more effective modern SOF operator [17]. For this integration to succeed, SOF operators must receive structured and ongoing training that bridges the gap between cyber specialists and traditional combat-focused personnel. Without integrated training

environments, SOF may miss opportunities to leverage cyber tools effectively, which could diminish their operational effectiveness in high-stakes scenarios [19]. Thus, a robust and dynamic training program that incorporates both cyber and traditional warfare skills is essential for developing multi-domain operators who can respond to contemporary challenges in the digital age.

Training for the Cyber-SOF Operator

To enable SOF personnel to conduct cyber operations effectively, specialized training in cybersecurity, digital forensics, network penetration, and cyber threat analysis is crucial. Training in network defense, vulnerability analysis, and digital forensics enables SOF operators to conduct both defensive and offensive operations in cyberspace [19]. As cyber threats become more sophisticated, operators must be capable of identifying vulnerabilities and responding to cyber threats in real time. Digital forensics is particularly vital for tracking adversary activities [20], while also stressing the importance of penetration testing and threat detection as key components of effective cyber operations [17]. This suggests that the implementation of a cyber training program adopted into the SOF enterprise as a whole and at the specific SOF unit level. That training program should incorporate unit funds to provide the study materials for a variety of cyber/information technology (IT) focused certifications starting with the basic fundamentals. A proposed escalation of certification would be:

Figure 2: Certification Progression

Governing Body	Certification	Outcome of that Certification
CompTIA	A+	Outcomes include validating IT skills, enhancing troubleshooting abilities, and preparing for roles in hardware, software, networking, and IT support environments.
CompTIA	Network +	Outcomes include validating networking knowledge, improving troubleshooting skills, and preparing individuals for roles in network administration, security, and IT infrastructure management.
CompTIA	Security +	Outcomes include validating cybersecurity knowledge, enhancing threat management skills, and preparing individuals for roles in network security, risk management, and security administration.
CompTIA	CompTIA Advanced Security Practitioner + (CASP+)	Outcomes include demonstrating advanced security knowledge, preparing for leadership roles in enterprise security, and validating skills in risk management, cryptography, and security architecture.
EC-Council	Certified Ethical Hacker (CEH)	Outcomes include mastering ethical hacking techniques, understanding vulnerabilities, and gaining skills to assess and secure network systems against cyber threats.
OffSec	Offensive Security Certified Professional	Outcomes include developing practical penetration testing skills, mastering exploitation techniques, and gaining hands-on experience to identify and mitigate vulnerabilities in network systems.

In addition to the certification progression path, evolving deployment exercises (Marine Special Operations Command – MARSOC) Raven exercise to include cyber relevant exercises illustrated in the following figure:

Figure 3: Cyber Exercise Integration into Deployment Exercises

Exercise	Scenario	Objective	Focus Area
Simulated Cyberattack on Critical Infrastructure	A cyberattack targets key communication, logistics, or energy infrastructure in a simulated hostile environment.	Train SOF units in identifying, mitigating, and responding to cyberattacks while maintaining operational effectiveness.	Incident response, network security, communication disruption, and system recovery.
Phishing and Social Engineering Attack Simulation	Cyber adversaries attempt to gather intelligence through spear-phishing emails or phone calls aimed at SOF personnel	Teach SOF personnel how to recognize and respond to social engineering attacks that might compromise operational security	Awareness training, email and communication security, and response protocols
Red Team vs. Blue Team Cyber Defense	Red team (attackers) simulate a cyber breach attempting to compromise SOF operational systems, while the Blue team (defenders) must detect and neutralize the threat	Test SOF's cyber defense capabilities in real-time during a live exercise to evaluate the security of communication systems, command control, and logistics infrastructure	Incident detection, response time, vulnerability assessment, and network defense
Cyber-Physical Systems Attack (Simulated)	A cyberattack compromises physical systems like drones, robotics, or automated vehicles used in SOF operations	Assess SOF's ability to recognize and respond to the impact of a cyberattack on physical equipment crucial to mission success	Integration of cybersecurity with physical security, equipment recovery, and continuity of operations
Data Exfiltration and Information Warfare	Adversaries attempt to infiltrate and exfiltrate sensitive data from SOF's communication systems or intelligence networks	Train SOF in securing sensitive data, maintaining operational security, and preventing the unauthorized exfiltration of mission-critical information	Data protection, encryption, secure communication, and operational secrecy

(Cont'd.)

Exercise	Scenario	Objective	Focus Area
Simulated Advanced Persistent Threat (APT) Response	A sophisticated, long-term cyberattack that infiltrates SOF systems, with the goal of stealing sensitive information over time without immediate detection	Develop SOF's capacity for long-term cybersecurity resilience and response to stealthy, persistent adversary actions	Intrusion detection, network monitoring, incident response, and system restoration.
Crisis Management in Cyber Warfare	A major cyber incident occurs during a SOF operation, such as a denial-of-service attack that disrupts mission communications or a malware infection in command systems	Simulate decision-making under stress, focusing on crisis management and recovery during an ongoing cyberattack while conducting critical missions	Command-and-control continuity, decision-making under pressure, multi-domain operations, and interagency collaboration
Cyber Threat Intelligence Sharing and Collaboration	SOF units must collaborate with other military or intelligence agencies to share cyber threat intelligence in real-time during a deployment	Enhance SOF's interoperability and effectiveness by working with other agencies to analyze and respond to emerging cyber threats	Intelligence sharing, threat analysis, interagency communication, and situational awareness

With the integration of these cyber capabilities, SOF operators will not only enhance their offensive and defensive cyber skills but also gain a technological edge over adversaries as well as increase their knowledge and awareness of how cyber integrates into traditional SOF and conventional operations. The continuous evolution of cyber threats necessitates ongoing training to ensure that SOF units can maintain a technological advantage and effectively respond to the dynamic nature of cyber warfare.

Redefining the Role of SOF in a Cyber-Physical Environment

As warfare becomes increasingly cyber-physical, SOF must adapt to a landscape where both traditional military operations and cyber capabilities are seamlessly integrated. This shift requires a redefinition of the SOF operator's role to encompass both kinetic and cyber operations. This suggests that SOF units must master cyber operations to effectively disrupt adversary networks, secure digital infrastructures, and execute complex operations across domains [21]. The integration of cyber capabilities into SOF operations allows for enhanced mission precision and reduced risk. For example, cyber operations such as disabling enemy communications can complement traditional kinetic strikes, creating a more effective multi-domain operation. Additionally, digital foot-printing through passive and active reconnaissance allows SOF teams to have a better understanding of their battle space and who (the enemy or potential enemy) is operating within their battlespace giving them the ability to have greater efficiency in mission planning and resource allocation as well as targeting precision.

The emergence of the "cyber-SOF operator" reflects the growing necessity for SOF personnel to function adeptly in both physical and virtual spaces. These operators must possess proficiency in cyber forensics, network penetration, and real-time intelligence gathering, while also mastering conventional warfare skills. The fusion of these skill sets allows SOF operators to engage in a wide array of operations, from disrupting critical infrastructure to conducting intelligence operations in cyberspace [20] [16]. Accentuating the convergence of cyber and physical domains will define the future of warfare, and as such, cyber-SOF operators will be crucial in maintaining SOF's operational superiority across all environments [19].

This implies that SOF must adapt to the growing integration of cyber and physical warfare, as both domains become increasingly interconnected. The future of warfare will require SOF to leverage cyber capabilities alongside traditional military tactics to maintain operational superiority.

Collaboration Across Domains and Sectors

The integration of SOF teams with cyber specialists, intelligence units, and non-military entities such as the NSA, private cybersecurity firms, and academic institutions is essential for ensuring mission success in the increasingly digital battlefield. Cyber specialists enable SOF to protect critical infrastructure, disrupt adversary networks, and gather actionable intelligence [22]. Intelligence units further enhance mission effectiveness by providing insights into enemy vulnerabilities, guiding SOF operations with targeted and strategic precision.

Collaborative operations with private sector cybersecurity firms, government agencies, and academia offer SOF access to cutting-edge technologies, threat intelligence, and specialized expertise that can significantly improve operational outcomes. The NSA's role in cyber intelligence and network defense provides SOF with crucial information regarding adversary capabilities [23]. These partnerships are instrumental in navigating sophisticated cyber threats and integrating cyber capabilities into traditional SOF missions, ultimately increasing mission success rates in complex and contested environments.

Adapting to the Pace of Technological Change

The rapid pace of technological change and the sophistication of cyber threats require SOF to prioritize continuous learning. The argument can be made that ongoing education and training are critical for staying ahead of emerging cyber threats [24]. Collaborating with academic institutions and tech companies, specifically smaller start-up companies, allows military personnel to access the latest research, innovations, and cybersecurity strategies, thus fostering a dynamic knowledge exchange that can be applied to SOF operations. Furthermore, simulation-based training exercises [25], are invaluable in preparing SOF personnel for the unpredictable nature of cyber-attacks. Lastly, it can be suggested the Joint Special Operations University (JSOU) expand their cyber education to include tactical level cyber training and awareness therefore being able to reach the SOF enterprise from a point of academic authority.

The integration of emerging technologies such as artificial intelligence (AI), machine learning (ML), and quantum computing will further shape SOF's ability to conduct cyber operations. AI and ML can facilitate the rapid analysis of vast amounts of data, enabling SOF to detect potential threats and automate responses in real time [26]. Quantum computing promises to revolutionize cybersecurity by providing advanced encryption methods, thereby safeguarding mission-critical communications [27]. As these technologies mature, SOF will need to adapt their training programs to incorporate these advancements, ensuring that they remain at the forefront of cyber warfare strategies.

The evolving role of Special Operations Forces in the cyber-dominated battlefield highlights the necessity of integrating cyber capabilities into SOF missions. As warfare increasingly converges with the digital domain, SOF must adapt by developing proficiency in both traditional combat skills and cyber operations. Through specialized training, cross-domain collaboration, and continuous learning, SOF units can remain agile, responsive, and effective in countering modern threats. The future of SOF will depend on their ability to operate seamlessly across both the physical and virtual spaces, ensuring mission success in the complex and interconnected battlefield of the 21st century.

Conclusion

The integration of cyber warfare capabilities into Special Operations Forces (SOF) operations has transformed military engagement, reshaping both strategic and tactical approaches in modern conflict. The increasing reliance on digital tools by adversaries compels SOF to adapt by enhancing their cyber expertise, incorporating advanced technologies, and ensuring operational flexibility. SOF's future role in joint and combined cyber operations will be pivotal, with these forces becoming central to rapid-response scenarios and strategic cyber engagements. The incorporation of cyber capabilities into SOF's operational framework ensures enhanced precision, agility, and resilience, particularly in complex multi-domain operations. As cyber threats continue to evolve, SOF's ability to seamlessly integrate offensive cyber operations with traditional kinetic tactics will amplify their effectiveness, enabling them to gain and sustain a strategic advantage, especially when facing near-peer adversaries [28] [29]. Additionally, understanding the legal and ethical considerations of offensive cyber operations, such as compliance with international law and addressing attribution challenges, remains critical to maintaining legitimacy and avoiding escalation in high-stakes engagements [29] [30]. Consequently, the future of SOF in cyber operations depends on the continuous evolution of their cyber capabilities, collaboration with conventional forces, and a deep understanding of the technological, legal, and strategic complexities inherent in digital warfare.

Importance of Integrating Cyber Capabilities with Traditional SOF Missions

The integration of cyber capabilities with traditional SOF missions is essential for enhancing operational effectiveness and adaptability in contemporary warfare. By merging cyber warfare with conventional kinetic operations, SOF units gain the ability to leverage non-kinetic methods to disrupt adversary command and control, communications, and decision-making processes. This strategic integration significantly enhances mission success while minimizing collateral damage and reducing

operational risks. The dynamic role of cyber operations in SOF missions provides increased flexibility, offering a critical advantage in multi-domain environments [29]. Moreover, the use of cyber tactics allows SOF to target adversary assets with precision, achieving disruption without resorting to traditional warfare methods, thereby reducing the likelihood of escalation (gray zone warfare) [28]. This convergence fosters strategic advantages, allowing SOF to operate as effectively in cyberspace as on the battlefield, ensuring that operations are synchronized across domains to counter modern adversary capabilities—particularly in cyber-dominant environments [30]. The strategic value of this integration is indispensable in advancing SOF's ability to meet the challenges of future warfare, especially in near-peer conflicts where technological and cyber dominance will be crucial [29]. This suggests that integrating cyber operations into SOF missions provides significant advantages by enhancing flexibility and precision in multi-domain environments. Cyber tactics enable SOF to disrupt adversary assets without escalating to traditional warfare, particularly in gray zone warfare, where confrontation is more subtle. This convergence allows SOF to operate effectively in both cyberspace and physical battlefields, ensuring synchronization across domains to counter adversaries' cyber capabilities. As technology and cyber dominance become increasingly vital in future conflicts, particularly against near-peer rivals, the integration of cyber operations into SOF strategies will be essential for maintaining operational superiority and adapting to the evolving landscape of modern warfare.

Implications for Military Strategy and Policy

The changing nature of cyber warfare will significantly shape future military doctrines, requiring substantial modifications to traditional strategies and policies. As cyber threats evolve in sophistication and scope, military doctrine must incorporate strategies to address the unique challenges posed by the digital domain. Future military operations will necessitate a blend of kinetic and non-kinetic strategies, with cyber warfare emerging as a core component of military engagements [29]. This shift will require military forces to adapt their strategic objectives, recognizing that cyberattacks can now cripple adversary infrastructure with the same, if not more, significance as traditional warfare. Posing the argument that the strategic flexibility provided by cyber operations allows military forces to conduct precise actions with minimal risk, necessitating an agile approach to planning and execution [28]. This suggests that cyber operations enable military forces to conduct highly precise actions with minimal risk. This requires an agile approach to planning and execution, emphasizing flexibility, speed, and adaptability in military strategy. The use of cyber operations can allow forces to achieve specific objectives with reduced physical engagement, lower risks, and fewer consequences for the broader operational environment suggesting that

the infusion of cyber is necessary for defeating adversary information operation campaigns. Furthermore, incorporating cyber capabilities into military doctrine requires the development of policies that address the ethical, legal, and operational complexities of cyber warfare, especially concerning issues of attribution and escalation [30]. As the role of cyber tools in military operations expands, the development of comprehensive doctrine that integrates both digital and conventional warfare will become increasingly critical. This will require changes in military training, legal frameworks, and international cooperation to ensure readiness for future conflicts [29]. These shifts will ultimately determine how military organizations adapt to the technological landscape and ensure that cyber warfare is fully integrated into future operations.

Policy Recommendations for SOF Cyber Training, Resource Allocation, and Joint Cyber Operations

To enhance the capabilities of Special Operations Forces (SOF) in the cyber domain, several key policy recommendations are vital. First, comprehensive training programs that integrate cyber expertise directly into SOF operations are essential. The value of simulation-based training, which prepares operators for cyber threats while enhancing decision-making skills in high-stress situations [25]. Moreover, investments in advanced technologies such as artificial intelligence (AI) and machine learning (ML) are crucial to augmenting SOF's cyber capabilities. These technologies enable real-time threat detection and autonomous responses, improving operational efficiency and mitigating risks [26]. Furthermore, prioritizing joint cyber operations with other military branches and intergovernmental partners will foster a more robust and collaborative defense framework. Such partnerships improve situational awareness and resource sharing, strengthening military cyber defenses [31]. Allocating resources to joint exercises and coordinated cyber response teams will ensure SOF units are equipped to respond to increasingly complex cyber threats. As the cyber domain continues to evolve, these policy measures will be essential to maintaining SOF's strategic advantage and readiness.

Future Research Directions

Future scholarly investigations into the impact of cyber warfare on SOF and other military units should explore the long-term effects of cyber capabilities on conventional military strategies. One area of focus could be the development of adaptive cyber defense systems tailored to the decentralized nature of SOF operations [32]. Additionally, the intersection of cyber warfare and psychological operations offers intriguing possibilities for understanding how cyber tools can influence adversary decision-making processes [33]. Exploring the application of AI and machine learning within cyber operations could further enhance SOF's predictive capabilities, preparing them for unpre-

dictable cyber threats [34]. This would also foster a potential study of the SOF enterprise's understanding of the application of AI at the strategic, operational, and tactical level. Additionally, researching the integration of cross-domain cyber-attacks and their real-time application could greatly improve SOF's efficiency in cyber-driven warfare.

Challenges and Opportunities for SOF in the Evolving Cyber Landscape

As the cyber domain continues to evolve, SOF operators will face both challenges and opportunities in incorporating cyber capabilities into their missions. The sophistication of emerging threats, such as advanced persistent threats (APTs) and cyber-enabled disinformation campaigns, presents significant challenges to mission integrity and security [34]. However, the increasing reliance on digital infrastructure provides opportunities to enhance cyber defense skills and ensure operational effectiveness in hostile environments [32]. Additionally, the decentralization of operations, while offering flexibility, introduces new vulnerabilities that adversaries can exploit. Yet, opportunities exist in leveraging AI and machine learning to improve predictive cyber defense capabilities, thereby enhancing real-time decision-making and response mechanisms [24]. This implies that there is potential to use artificial intelligence (AI) and machine learning (ML) to significantly enhance cybersecurity. By applying AI and ML, SOF units could improve their ability to predict cyber threats and attacks before they occur. These technologies can help analyze large volumes of data in real-time, enabling quicker identification of patterns, vulnerabilities, and anomalies.

This, in turn, enhances decision-making and allows for faster, more effective responses to potential cyber threats, strengthening overall defense mechanisms and minimizing damage from cyber incidents. Finally, research into adaptive cyber defense strategies will provide SOF with the tools needed to maintain resilience in increasingly dynamic and unpredictable environments [32]. This suggests that conducting research into adaptive cyber defense strategies will equip Special Operations Forces (SOF) with advanced tools and techniques to effectively protect their operations in complex and rapidly changing environments. By focusing on adaptability, SOF will be able to quickly adjust to evolving cyber threats, ensuring continued operational resilience and security despite unpredictable challenges.

The future of warfare will increasingly involve complex interactions between cyber and physical domains. For SOF, this presents both challenges and opportunities. As cyber threats continue to evolve, SOF must integrate advanced cyber capabilities into their operations to maintain relevance and strategic advantage. The convergence of cyber and kinetic warfare will define future military engagements, and SOF must evolve to meet these demands. By focusing on continuous training, interdisciplinary collaboration, and the integration of cutting-edge technologies, SOF will remain agile and effective in navigating the complexities of modern warfare. The successful integration of cyber capabilities will ensure that SOF are prepared for the rapidly changing landscape of global conflict, enhancing their operational effectiveness and resilience in an increasingly interconnected world.

AJ Rutherford is a retired Marine Special Operations Team Chief currently serving as an overseas security consultant in the private sector and holds a PhD in Information Security and Assurance and an MA in Strategic Intelligence. His primary area of research includes integration of technology, specifically artificial intelligence (AI) into Special Operations. Highlights of his research include what SOF looks like in fifth generation warfare. Future research addresses integrating Large Language Models (LLMs) and Natural Language Process (NLP) into SOF decision making and combat planning as well as conducting a quantitative study on SOF assessing the tactical level operator's aptitude in Python Programming in order to effectively manipulate Structured Query Language (SQL) databases in order to integrate AI into mission planning. He welcomes opportunities for continued research, collaboration, and presentation.

References

[1] Bacevich, A. J. (2013). The new American militarism: How Americans are seduced by war. Oxford University Press.

[2] Garamone, J. (2017). Special operations forces: Key role in counterinsurgency. Defense.gov.

[3] Nakasone, P. M. (2023). *2023 posture statement of General Paul M. Nakasone, Commander, United States Cyber Command*. U.S. Cyber Command.

[4] Libicki, M. C. (2021). Cyber War and Peace: The Strategic Dimensions of Digital Conflict. Washington, D.C.: Georgetown University Press.

[5] Kostyuk, N., & Zhukov, Y. M. (2022). Invisible Digital Fronts: The Role of Cyber Operations in Contemporary Warfare. *Journal of Conflict Resolution, 66*(3), 567-595.

[6] Kilcullen, D. (2006). Counterinsurgency. Oxford University Press.

[7] Liddell Hart, B. H. (1991). Strategy: The Indirect Approach. Faber & Faber.

[8] Trinquier, R. (1964). Modern Warfare: A French View of Counterinsurgency. Praeger Publishers.

[9] Jordan, A. (2010). The Evolution of Asymmetric Warfare: Implications for Military Strategy. *Journal of Strategic Studies*, 33(4), 423-442.

[10] Libicki, M. C. (2020). Cyberspace in peace and war. Georgetown University Press.

[11] Nye, J. S. (2021). The future of power. Public Affairs.

[12] Sullivan, K. (2019). Cybersecurity and Cyberwar: What Everyone Needs to Know. Oxford University Press.

[13] Rid, T. (2012). Cyber in the shadows: Why the future of cyber operations will be covert. *Journal of Strategic Studies*, 35(1), 1–28.

[14] Zetter, K. (2017, January 10). The Ukrainian power grid was hacked again. Wired. https://www.wired.com/2017/01/ukrainian-power-grid-hacked-again/

[15] Booz Allen Hamilton. (2016). *The future of warfighting: Cyber enabling con-*

vergence. Booz Allen Hamilton. Retrieved from https://www.boozallen.com/

[16] Duggan, P. M. (2016, January 8). Why Special Operations Forces in US Cyber-Warfare? *The Cyber Defense Review*.

[17] Beaurpere, W., & Marsh, N. (2024). Space, Cyber, and Special Operations: An influence triad for global campaigning. Modern War Institute.

[18] Fetters, R., Hickman, B., Jones, R., & Reed, J. (2024, August 9). Winning the first fight: Experimenting with Army Special Operations Forces' contributions in large-scale combat operations. Modern War Institute.

[19] Golding, J. (2022, November 11). Byte, with, and through: How Special Operations and Cyber Command can support each other. War on the Rocks.

[20] Van Hooren, J. (2019). The imperative symbiotic relationship between SOF and cyber: How Dutch Special Operations Forces can support cyber operations [Thesis, Netherlands Ministry of Defense]. Naval Postgraduate School.

[21] Starling, C. G., & Marine, A. (2024, March 7). Stealth, speed, and adaptability: The role of special operations forces in strategic competition. Atlantic Council.

[22] Lamb, C. (2021). Cyber-SOF integration for mission success. Military Operations Review.

[23] Sadowski, J. (2020). Building cyber partnerships: Leveraging private sector expertise for military operations. *Cybersecurity Journal*.

[24] Williams, D., & Anderson, C. (2020). Cyber defense strategies in military operations: The need for continuous learning. *Journal of Military Cybersecurity*, 50(3), 112-123.

[25] Taylor, R., & Evans, P. (2023). Advancing cyber readiness in military forces: The role of simulations in continuous learning. *Journal of Cyber Warfare and Security*, 43(1), 88-102.

[26] Jensen, K., & Patel, R. (2022). Artificial intelligence in military cyber operations: A new frontier. *Cybersecurity and Operations Review*, 41(3), 98-110.

[27] Williams, D. (2021). Quantum computing and the future of cryptographic defense in military operations. *Journal of Digital Security*, 34(4), 45-56.

[28] Brown, D., & Reed, P. (2020). Strategic flexibility in SOF through cyber oper-

ations. *Journal of Military Strategy and Operations,* 22(4), 112-125.

[29] Fitzgerald, M., & Andersson, S. (2023). Cyber warfare and state sovereignty: Legal implications for military operations. *Journal of Military Law and Ethics,* 41(2), 134-149.

[30] Schmitt, M. N. (2021). The legality of cyber warfare: A framework for the future. *International Review of the Red Cross,* 102(1), 34-48.

[31] Graham, S., & Lee, M. (2021). Integrating advanced technologies into military cyber operations: Opportunities and challenges. *Cyber Defense Review,* 39(2), 80-92.

[32] Voss, M., & McNamara, S. (2023). Adaptive cyber defense for decentralized forces: Strengthening SOF's resilience. *Cybersecurity in Defense,* 19(1), 77-91.

[33] Brennan, P., & Garcia, L. (2023). Cyber warfare and psychological operations: The role of cyber in influencing adversary decision-making. *Journal of Military Cyber Operations,* 54(2), 45-58.

[34] Smith, R., & Carter, J. (2023). Artificial intelligence in military cyber operations: Enhancing predictive capabilities. *Military Technology Review,* 40(3), 112-125.

Silent Sacrifices: Political Predictors of UN Peacekeeper Fatalities

Kathryn M. Lambert[1]

Abstract

The United Nations reports 4,348 peacekeeper fatalities during mission deployments from 1948 through 2023. However, peacekeeper security receives limited attention in the literature. Forty-seven United Nations peacekeeping missions between 1991 and 2019 were studied to address this gap. The Peacekeeping Mandates Dataset (PEMA) was used to identify peacekeeper tasks during randomly selected event years. PEMA also classifies whether the peacekeepers assisted, monitored, or secured each task. This study uses the tasks classified in the assist modality to develop three models that reflect varying mission intentions: 1) the State Model includes missions where peacekeepers assisted the state in maintaining or extending control over territory and when peacekeepers assisted in the reduction of conflict and the building of community confidence; 2) the Reform Model includes missions where the peacekeepers assisted in reforming law enforcement institutions and engaged in child protection duties; and 3) the Military Model includes missions where the peacekeepers assumed control over small arms and light weapons and moved troops and heavy weapons out of specific locations. The study contributes to the existing literature by finding that only when peacekeepers assume control of small arms does their likelihood of fatalities increase.

Keywords: Peacekeeping Missions; Peacekeeper Security; Peacekeeping Mandates

Sacrificios silenciosos: predictores políticos de las muertes de las fuerzas de paz de la ONU

Resumen

Las Naciones Unidas informan 4.348 muertes de personal de mantenimiento de la paz durante despliegues de misión desde 1948 hasta 2023. Sin embargo, la seguridad del personal de mantenimiento de

1   Kathryn.Lambert@mycampus.apus.edu

doi: 10.18278/si.10.1.5

la paz recibe poca atención en la literatura. Cuarenta y siete misiones de mantenimiento de la paz de las Naciones Unidas entre 1991 y 2019 se estudiaron para abordar esta brecha. El Conjunto de datos de mandatos de mantenimiento de la paz (PEMA) se utilizó para identificar las tareas de mantenimiento de la paz durante años de eventos seleccionados aleatoriamente. PEMA también clasifica si el personal de mantenimiento de la paz asistió, monitoreó o aseguró cada tarea. Este estudio utiliza las tareas clasificadas en la modalidad de asistencia para desarrollar tres modelos que reflejan diversas intenciones de misión: 1) el Modelo de Estado incluye misiones en las que el personal de mantenimiento de la paz ayudó al estado a mantener o extender el control sobre el territorio y cuando el personal de mantenimiento de la paz ayudó a la reducción del conflicto y al fomento de la confianza de la comunidad; 2) el Modelo de Reforma incluye misiones en las que el personal de mantenimiento de la paz ayudó a reformar las instituciones policiales y participó en tareas de protección infantil; y 3) el Modelo Militar incluye misiones en las que las fuerzas de paz asumieron el control de armas pequeñas y ligeras y trasladaron tropas y armas pesadas de ubicaciones específicas. El estudio contribuye a la literatura existente al concluir que solo cuando las fuerzas de paz asumen el control de armas pequeñas aumenta la probabilidad de muertes.

Palabras clave: Misiones de mantenimiento de la paz; Seguridad de las fuerzas de paz; Mandatos de mantenimiento de la paz

无声的牺牲：联合国维和人员伤亡的政治预测因素

摘要

据联合国报告，自1948年至2023年，维和人员在任务部署期间共有4348人死亡。然而，鲜有研究聚焦于维和人员的安全。为了弥补这一空白，研究了1991年至2019年期间47个联合国维和特派团。维和授权数据集(PEMA)被用于识别随机选择的事件年份中的维和人员任务。PEMA还对维和人员是否协助、监督或保障每项任务一事进行了分类。本研究使用按协助模式分类的任务，提出了三个反映不同特派团意图的模型：1)国家模型，包括维和人员协助国家维持或扩大对领土的控制以及维和人员协助减少冲突和建立社区信任；2)改革模型，包括维和人员协助改革执法机构并参与儿童保护职责；3)军事模式，包括维和人员接管轻小型武器，并将部队和重型武器调离特定地点。本研究发现，只有当维和人员接管轻小型武器时，其伤亡概率才会增加，这为现有文献作贡献。

关键词：维和任务，维和人员安全，维和授权

In May 2023, during a celebratory speech recognizing the 75th anniversary of peacekeeping, Secretary-General António Guterres described peacekeepers as "the beating heart of our commitment to a more peaceful world"" (UN press release, 2023). Seventy-five years earlier, on May 29, 1948, the Security Council adopted Resolution 50, calling for a cessation of hostilities in Palestine and authorizing a United Nations (UN) mediator to monitor the ceasefire with the assistance of military observers. This action launched the United Nations' peacekeeping enterprise. Less than ten years later, the organization authorized the deployment of its first armed mission in response to the Suez Crisis. The scope and depth of its peacekeeping missions expanded over subsequent decades. As of May 2024, the United Nations authorized seventy-one peacekeeping missions to promote international peace and security. However, as Robert Kayinamura, Deputy Permanent Representative of the Republic of Rwanda to the United Nations, noted, "Peacekeepers are deployed in places where there is no peace to keep" (2022, p. 2). The often violent and tumultuous environment in which peacekeepers operate leads inevitably to fatalities. The United Nations reports 4,348 peacekeeper fatalities in mission deployments from 1948 through the end of 2023. The organization honors the sacrifices of its peacekeepers in ceremonial and symbolic actions. In 2002, the UN General Assembly officially declared May 29 the International Day of UN Peacekeepers. This annual observance, dedicated to the contributions of peacekeepers, commemorates those who gave their lives in the service of peace. Moreover, within the United Nations grounds, a memorial site serves as a place to pay tribute to peacekeepers killed in the line of duty. Pursuing and maintaining peace and security takes a toll on those entrusted with these responsibilities.

The United Nations protects its personnel using a robust Security Risk Management framework. However, it was not until 2008 that the UN published its first codification of the lessons learned during its peacekeeping operations. The *United Nations Peacekeeping Operations: Principles and Guidelines* identifies various criteria the Security Council may consider before deploying peacekeepers. Among those issues is the need to "reasonably ensure" the safety and security of UN personnel (p. 48). This publication discusses the UN security management system for its international and national staff and its responsibilities as the head of the military and police components of peacekeeping forces "to ensure that the best possible security arrangements are put in place for all personnel" (p. 80). In 2022, the Secretary-General issued a report on the security of UN personnel pursuant to General Assembly Resolution 76/127, which calls for enhanced resources for its Department of Safety and Security. While the framework mentioned above affords a broad range of security measures and expertise to UN missions, the organization's Security Policy Manual identifies host governments as primarily responsible for securing UN personnel with the UN, acting in the capacity of an employer, playing a supporting role (2017, p. 8).

The United Nations' efforts to formulate security protocols to improve the safety of its personnel must be an ongoing endeavor as threats and response capabilities evolve. Scholarly research on peacekeeper security can contribute to this effort. Unfortunately, limited research exists on the predictors of peacekeeper fatalities. This study addresses the gap by posing the following research question: What are the political predictors of peacekeeper fatalities resulting from malicious acts? By identifying predictors of peacekeeper fatalities, this research will contribute valuable insights for enhancing the safety and security of peacekeepers as they seek to implement their mandates in often hostile and chaotic environments. A better understanding of peacekeepers' vulnerability to attack may contribute to the efficacy of peacekeeping missions.

Literature Review

Though the literature on peacekeeper security is scant, it covers two main points: legal protections afforded to peacekeepers and their vulnerability to attack. The legal protections surrounding peacekeepers reflect that provided to civilians. In other words, international law offers the same protection to peacekeepers as it does to civilians as long as the peacekeepers are not acting as belligerents in the conflict. Bangura's (2010) work explores the conditions under which peacekeepers qualify as belligerents and the impact on treaty protections when the peacekeeper operates outside the scope of the mandate. The 1994 Convention on the Safety of United Nations and Associated Personnel represents a step in creating specific protections for peacekeepers to enhance their security. Shortly after the passage of this treaty, legal experts weighed in on its effectiveness. In a 1995 article, Bloom argued that the increase in peacekeeping missions and corresponding attacks against those missions resulted in states' reluctance to staff the operations because of the dangerous environment posed to their nationals serving as peacekeepers. Bloom believed the treaty would address this concern and increase states' mission participation (p. 630). Also publishing after passage of the1994 treaty was Greenwood (1996), who took a more pessimistic view of the treaty's utility by expressing concern that the added protection to peacekeepers as a class would make it more difficult for them to claim protections provided to belligerents in the 1949 Geneva Conventions (p. 186). Only two years later, the 1998 Rome Statute classified attacks against peacekeepers as war crimes. Gadler (2010) explores the application of this statute in prosecutions of peacekeeper attacks by the International Criminal Court and an international tribunal and special court. The author concluded that using the statute in these types of prosecutions led to broadening the types of crimes prosecuted overall with the law.

A second focus of the literature on peacekeeper security looks at vulnerability to attack as a function of the distribution of power among the belligerents. Two articles were published in 2013; the one by Salverda lamented the lack of research on peacekeeper security. Salverda asked why rebels attack peacekeepers in some cases while not in others and concluded that when rebels are the stronger party, the peacekeepers protect the weaker party in the conflict, which in this case is the state. The study found that the stronger the rebels become, the greater the probability that they will attack peacekeepers. Ruggeri et al., who published the second article in 2013, examined peacekeeping operations in African civil wars between 1989 and 2005. These authors found that the size of the mission reflected the resolve of the United Nations, which enticed cooperation from the rebels. The stronger the government, the greater the cooperation of the rebels with the peacekeepers (Ruggeri et al., 2013, p. 389). A final work addressing violence against peacekeepers, written by Fjelde et al. in 2016, focused on violent attacks in sub-Saharan Africa from 1989

to 2009. This study found that as rebels suffer battlefield losses, violent attacks against peacekeepers increase (p. 611). The authors explain that as rebels incur losses within a peacekeeping operation, the rebels deem the mission a disadvantage. Peacekeepers face even greater risk when militarily aligned with government forces. An early work by Seet and Burnham (1998) attributed the rise in peacekeeper fatalities to the expanding nature of peacekeeping operations. The authors found that the risk of attacks increases with assignment to Africa, involvement in a peace enforcement mission, and participation in a humanitarian effort.

Two other studies shed light on the conditions under which peacekeepers are attacked, though neither deals directly with violence against peacekeepers. First is a work by Duursma in 2019, which examined the use of obstruction and intimidation by armed actors against peacekeepers. The study found that when the armed actors attacked civilians, it prompted their resistance to the peacekeepers. The author believes that this tactic is an attempt by the rebels to prevent the peacekeepers from using violence to protect civilians and to discourage human rights investigations. The second work is a 2024 study by Abbs and Duursma that addressed the issue of whether UN peacekeepers, once deployed, remain close to their base or venture into the violent areas where they are most needed. The study concluded that peacekeepers patrol violent areas, suggesting they often put themselves in the most volatile environments once deployed.

Since 1948, over two million men and women have served as peacekeepers. However, scholarly attention to peacekeeper security remains lacking despite the increasing importance of international peacekeeping missions and the escalating risks peacekeepers face in conflict zones. This study aims to fill a noticeable gap in the literature by identifying political predictors of peacekeeper fatalities and offering valuable insights into the underlying factors that contribute to the risks faced by peacekeepers.

Methodology

Data Description

The dataset includes 47 United Nations peacekeeping missions (listed in Appendix A) that operated at some point between 1991 and 2019. The 47 missions vary in size and duration and also include a broad range of activities and scope of operations. In order to ensure the independence of observations, a randomly selected year from each mission serves as the unit of analysis. The thin literature on peacekeeper security hinders the development of a theoretical framework. As a result, this study focuses primarily on the missions' tasks as the explanatory cause of violence against peacekeepers. Complementary variables to the mandated tasks were identified and subjected to univariate regression analysis to assess their utility in the study. Appendix B contains the list of 56 independent variables subjected to univariate regression analysis with their corresponding

coefficients. As the study aims to identify predictors rather than conduct hypothesis testing, a p-value of .10 rather than .05 was used (Ranganathan et al., 2017). The results of the analyses identified 15 independent variables that held predictive value for the dependent variable.

The Dependent Variable

The United Nations maintains data on the frequency and cause of fatalities during peacekeeping missions. The UN classifies fatalities as the result of malicious acts, accidents, illnesses, or other causes. As of June 30, 2023, the United Nations reports 2,967 total fatalities for the 47 missions in this study, with 734 (25%) the result of malicious acts. A dependent binary variable was created where the event year was coded (1) if fatalities caused by malicious acts occurred during the event year or (0) if no fatalities occurred during the event year or the fatalities resulted from causes other than malicious acts. The dataset represents nicely distributed data, with twenty event years (43%) containing maliciously motivated fatalities and twenty-seven event years (57%) that do not for a total of 47 cases.

The Independent Variables

Peacekeeping Mandates Dataset (PEMA)

The Peacekeeping Mandates Dataset (PEMA) was used to identify the tasks carried out in each event year in the study. PEMA codes 41 distinct tasks carried out by peacekeepers as outlined in mandates and subsequent updates. Tasks include a wide range of activities such as controlling small arms, protecting civilians, reforming the police or military, controlling the border, re-establishing government control, and strengthening democracy. The PEMA dataset uses three modalities to code whether peacekeepers monitor, assist, or secure each task. This study used data from the assist modality since it captures cases in which the peacekeepers actively engaged in the task. Therefore, the study excludes those tasks from PEMA's secure and monitor modalities.

Univariate regression analysis on PEMA's 41 tasks revealed the following tasks with a p-value of less than .10: 1) Control of Small Arms and Light Weapons – Peacekeepers assist in small arms and weapons collection, storage, and destruction programs; 2) Demilitarization – Peacekeepers assist in the withdrawal of troops and heavy weapons from specific areas; 3) State Authority Extension – Peacekeepers assist in re-establishing and extending government control over territory; 4) Local Reconciliation – Peacekeepers assist in local reconciliation tasks to strengthen trust and confidence in communities; 5) Child Rights – Peacekeepers assist in efforts to protect children during or after armed conflict; and, 6) Police Reform – Peacekeepers assist in reforming, restructuring, and rebuilding law enforcement institutions. Multicollinearity tests on the six PEMA variables reveal that the Variance Inflation Factors (VIF) for all six variables are below five, Tolerance Levels are all above .20, and the Condition Indices are all below 15, indicating that multicollinearity is not an issue among the six PEMA variables used in this study.

Other Independent Variables

In addition to the 41 PEMA tasks, 15 other variables were subjected to univariate regression analysis to assess their potential usefulness in the study. Of the 15 non-PEMA variables tested, nine produced p-values of less than .10. The study includes four of these nine variables: 1) Worldwide Governance Indicator for stability; 2) Number of uniformed UN personnel assigned to the mission, which includes military, police, and military advisors; 3) Level of civil liberties as reported by Freedom House; and, 4) Number of aid worker attacks during the event year identified by the Aid Worker Security Database (AWSD). Additionally, Uni Nexis searches identified aid worker attacks for years prior to 1997, when AWSD began collecting data.

The Three Models

Due to the small number of cases, logistic regression was used to test the three models. Each model reflects a unique purpose of the peacekeeping operation by assessing the influence of four independent variables on whether or not UN peacekeeper fatalities occurred during the event year. Each model contains two PEMA independent variables that showcase whether the peacekeepers' tasks influenced their vulnerability to attack, along with two other independent variables selected from among the significant variables in the univariate regression analysis.

Model 1: State Model

This model seeks to capture the peace-keeping mission pointedly designed to preserve state authority. The model relies on two PEMA variables to capture this mission feature. The first independent PEMA variable, state authority extension, includes those missions where the peacekeepers sought to maintain and extend state control over territory. The second independent PEMA variable, local reconciliation, includes those missions where peacekeepers sought to reduce conflict and build community confidence. Two other independent variables related to preserving state authority for the State Model were selected. First, the Worldwide Governance Indicator (WGI) of state stability was selected as a corresponding predictor of peacekeeper attacks, as higher levels of stability are likely to lead to state preservation as it generates political, economic, and social development. The second independent variable selected for the State Model is the number of UN uniformed personnel assigned to the mission during the event year. The task of the UN personnel in the State Model is to protect the state, which, in turn, may fuel animosity toward peacekeepers and increase the likelihood of their attack.

Model 2: Reform Model

The Reform Model includes those missions focused on reforming the state. Two PEMA independent variables were

selected to reflect this focus of the mission. The first variable includes those missions where peacekeepers reformed the state's law enforcement institutions. The second PEMA variable includes the missions where the peacekeepers actively engaged in child protection services. This model includes two other independent variables. First, the number of uniformed UN personnel assigned to the mission may predict peacekeeper fatalities, as the number of troops may reflect the level of commitment of the UN to reform. If the belligerents accept or tolerate the reform effort, the variable will hold no predictive value, but if the reform effort conflicts with the goals of the belligerents, then the variable will emerge as a predictor in the model. The second independent variable selected for the Reform Model reflects whether aid workers experienced attacks during the event year. Aid workers engage in humanitarian projects ranging from providing emergency relief to designing and constructing long-term sustainability projects. In the reform model, aid workers and peacekeepers strive to improve the well-being of those residing in the conflict zone. As a result, if belligerents attack aid workers, they likely have no regard for peacekeepers, so attacks against aid workers likely will predict attacks against peacekeepers in this model.

Model 3: Military Model

The focus of the Military Model is on peacekeeper control over military assets. This model uses the PEMA variable of control of small arms, which captures those missions in which the peacekeepers assume control over the weapons used by the belligerents. The second PEMA variable is demilitarization, which reflects the missions in which peacekeepers move troops and heavy weapons out of specific locations. The other two independent variables used in this model are the number of uniformed personnel assigned to the mission and the level of civil liberties in the state(s) where the mission operates. The number of UN personnel assigned to missions where the peacekeepers control and reduce military assets is expected to predict peacekeeper fatalities as the UN directly reduces the belligerents' military and political power.

Logistic Regression Results

Logistic regression was performed on each of the three models. All three models met the assumptions of logistic regression. First, the linearity of the continuous variables with respect to the logic of the dependent variable was assessed via the Box-Tidwell procedure. Based on this assessment, all continuous independent variables in each model were found to be linearly related to the logit of the dependent variable. Second, multicollinearity tests were conducted on each of the three models. The Variance Inflation Factors (VIF) for all variables are below five, the Tolerance Levels are above .20, and the Condition Indices are below 15, indicating that multicollinearity is not an issue in any models. Third, outliers were detected using the case diagnostic function in SPSS. There was one standard-

ized residual with a value of 2.775 in the State Model, two standardized residuals with values of 2.844 and 3.449 in the Reform Model, and two standardized residuals with values of 2.570 and 4.423 in the Military Model. A review of the outliers revealed that one case appeared in all three models while another case appeared in two models. The two cases were reviewed and kept in the analysis.

Model 1 – State Model

A binomial logistic regression was performed to ascertain the effects of four independent variables—state stability, the number of uniformed personnel assigned to a mission, a state authority mission task, and a local reconciliation mission task—on the likelihood of peacekeeper fatalities. The logistic regression model was statistically significant, $X^2(4)=25.15$, $p < .001$. The model explained 56% (Nagelkerke R^2) of the variance in peacekeeper fatalities and correctly classified 79% of cases. Sensitivity, or the percent of true positives in the model, was 70%, while specificity, or the percent of true negatives in the model, was 85.2%. The positive predictive value, or the percent of fatality cases predicted correctly, was 77.8%, and the negative predictive value, or the percent of cases without fatalities correctly predicted by the model, was 79.3%. As shown in Table 1, only two of the four predictor variables were statistically significant: stability and number of uniformed personnel assigned to the mission.

Table 1. *Logistic Regression Results for the State Model*

Variables in the Equation

		B	S.E.	Wald	df	Sig.	Exp(B)	95% C.I.for EXP(B) Lower	Upper
Step 1[a]	State_ Authority(1)	-.136	1.002	.019	1	.892	.872	.122	6.217
	Local_Reconcil. (1)	2.340	1.266	3.413	1	.065	10.376	.867	124.153
	Stability	-.106	.042	6.347	1	.012	.900	.829	.977
	Zscore: #Uniformed	1.180	.547	4.647	1	.031	3.255	1.113	9.517
	Constant	.835	.750	1.239	1	.266	2.304		

a. Variable(s) entered on step 1: State Authority, Local Reconciliation, Stability, Zscore: #Uniformed.

Increasing stability was associated with a reduction in the likelihood of peacekeeper fatalities. This result seems consistent with the general finding that peacekeeping missions reduce violence (see, for example, Ruggeri et al., 2017; Kim et al., 2020). Much of the effort that looks at the frequency of violence and peacekeeping operations focuses specifically on whether such missions reduce civilian casualties (see, for example, Bara, 2020; Fjelde, 2019; Haass and Ansorg, 2018). The authors of the PEMA dataset (Di Salvatore et al., 2022) note that nearly all UN peacekeeping mandates for the last two decades reflect the international norm of human security (p. 925). Lundgren et al. (2022) explain that protecting civilians has since become a criterion for judging the effectiveness of peacekeeping operations (p. 37). Most recently, Duursma et al. (2024) studied the impact of host-state consent and found that the quality of that agreement impacts the UN's protection of civilian activities.

The second significant variable in the model indicates that the larger the number of uniformed peacekeepers assigned to a mission, the greater the likelihood of peacekeeper fatalities. It seems that instability, representing a vulnerable and chaotic environment, may lead to the assignment of larger missions, which in turn increases the likelihood of peacekeeper fatalities. Di Salvatore (2019) notes that the UN does not randomly deploy peacekeeping missions but concentrates them in the most violent areas (p. 844).

The lack of significance of the two mandate variables—state authority extension and local reconciliation—deserves attention. Both of these tasks would likely impact the power distribution between the belligerents. The first task, state authority extension, deliberately bolsters state power. An enhancement of state power either reduces the political power of the insurgents or freezes them out of enhancing their power from controlling specific territory. Less direct but equally impactful on the insurgents is when peacekeepers engage in reconciliation efforts and seek to fortify communities in chaotic situations. Fjelde et al. explore the issue of power distribution in their 2016 study, which examines rebel losses to government forces in the context of peacekeeping missions. The authors conclude that rebels perceive themselves as disadvantaged in such a context and attack peacekeepers to restore their relative power. However, this study concludes that the loss of relative power, explicitly defined by these two tasks, does not emerge as a predictor of peacekeeper fatalities. This result seems more in line with the results found by Dorussen and Gizelis (2013), who found that rebels do not contest UN peacekeeper policies that strengthen the government. These authors speculate that such policies may offer new opportunities for cooperation among the actors.

Model 2 – Reform Model

A binomial logistic regression was performed to ascertain the effects of four independent variables—aid worker attacks, the number of uniformed personnel assigned to a mission, a task to reform law enforcement, and a mission task to protect children—on the likelihood of peacekeeper fatalities. The logistic regression model was statistically significant, $X^2(4)=18.48$, $p < .001$. The model explained 44% (Nagelkerke R^2) of the variance in peacekeeper fatalities and correctly classified 77% of cases. Sensitivity, or the percent of true positives in the model, was 65%, while specificity, or the percent of true negatives in the model, was 85.2%. The positive predictive value, or the percent of fatality cases predicted correctly, was 76.5%, and the negative predictive value, or the percent of cases without fatalities correctly predicted by the model, was 76.7%. As shown in Table 2, only two of the four predictor variables were statistically significant: Attacks against aid workers and the number of uniformed personnel assigned to the mission.

Table 2. *Logistic Regression Results for the Reform Model*

		B	S.E.	Wald	df	Sig.	Exp(B)	95% C.I.for EXP(B)	
								Lower	Upper
Step 1[a]	Police_ Reform(1)	.364	.839	.188	1	.665	1.439	.278	7.451
	Child_ Rights(1)	.673	.868	.602	1	.438	1.961	.358	10.748
	Aid Worker Victims(1)	1.791	.753	5.660	1	.017	5.993	1.371	26.201
	Zscore: #Uniformed	.853	.468	3.316	1	.069	2.346	.937	5.872
	Constant	-1.757	.821	4.581	1	.032	.173		

a. Variable(s) entered on step 1: Police Reform, Children's Rights, Aid Worker Victims, Zscore: #Uniformed.

The odds of peacekeeper fatalities are 5.993 times greater if aid workers operating in the same country during the same year as the peacekeepers were attacked. Scholars started to focus on the relationship between peacekeeping missions and attacks against aid workers due to the 2010 Brahimi Report, which expanded the role of peacekeeping missions. Concerns that the expanded role of peacekeepers would militarize or politicize aid (Hoelscher et al., 2017), shrink the humanitarian space (Sauter, 2022), or blur the lines between aid workers and others (Mitchell, 2015) abound. The body of work focused on the impact of peacekeeping missions on aid worker security. In 2015, Michell tested the blurred line theory by focusing on the Provisional Reconstruction Teams (PRTs) that operated in Afghanistan between 2010 and 2011. The blurred line theory posits that malicious actors cannot differentiate aid workers from non-aid workers in an environment where civilians and military actors work together. Mitchell (2015) found no evidence to support the blurred line theory in Afghanistan. In a later work, Mitchell and Kisangani (2021) conducted a large-scale study that considered the impact of the United Nations' integrated missions with a humanitarian support nexus in sixty-seven countries that experienced intrastate conflict between 1997 and 2018. The authors conclude in this study that integrated UN missions predict aid worker fatalities (p. 221). Alternatively, Sauter (2022) finds a blurred line between aid and non-aid workers and concludes that aid worker

attacks may decrease in an integrated mission due to a smaller humanitarian space rather than enhanced security (p. 644). The Reform Model shows that aid worker attacks predict peacekeeper fatalities when aid workers operate in the same state where peacekeepers buttress the state's effectiveness through reform. Attacks on peacekeepers aiming to reform the government, coupled with attacks against aid workers carrying out tasks associated with government duties, could signal insurgents' intentions to destabilize the state.

The number of uniformed personnel assigned to the mission was the second independent variable that proved significant. The Reform Model shows that the larger the number of uniformed personnel, the lower the likelihood of attacks against peacekeepers, suggesting that overwhelming force provides protection. Though Levin (2023) focused on aid worker security, his conclusion that large peacekeeping missions can better secure their environments and navigate hostile contexts also likely applies to peacekeeper security.

Neither PEMA variable—police reform and child rights—proved significant in the Reform Model. One study that dealt with the issue of reform in the literature was a 2017 study by Hoelscher, Miklian, and Nygård. These authors found that traditional mandates, such as observing a truce, result in more aid worker attacks, likely because the mission does not authorize using force. Alternatively, the authors found that transformational mandates, such as police, military, or judicial reform, lessened attacks against aid workers (p. 555). The results of this study are consistent with those of Hoelscher et al. (2017), as the police reform task did not predict peacekeeper fatalities.

Insight into the lack of significance for the protection of child rights may be found in the 2013 work by Dorussen and Gizelis. These authors found that both states and rebels partly resist UN peacekeeper policies that seek to improve human rights as neither party gains from the policy. Non-state actors, whether criminal or political, may recognize that attacking those who protect children could produce a backlash against the attackers both in their local community and from a regional or global audience. Additionally, attacking such peacekeepers may not align with the strategic goals of the attacker when other targets are available.

Model 3 – Military Model

A binomial logistic regression to ascertain the effects of four independent variables—the level of civil liberty in the state, the number of uniformed personnel assigned to a mission, a mission task where peacekeepers control small arms, and a mission task where troops and heavy weapons are removed from a specific area—on the likelihood of peacekeeper fatalities. The logistic regression model was statistically significant, $X^2(4)=27.83$, $p < .001$. The model explained 60% (Nagelkerke R^2) of the variance in peacekeeper fatalities and correctly classified 83% of cases. Sensitivity, or the percent of true positives in the model, was 75%, while specificity, or the percent of true negatives in the model, was 88.9%. The positive predictive value or the percent of fatality cases predicted correctly, was 83%, and the negative predictive value, or the percent of cases without fatalities correctly predicted by the model, was also 83%. As shown in Table 3, two of the four predictor variables proved statistically significant: UN control of small weapons and the level of liberty in the state.

Table 3. *Logistic Regression Results for the Military Model*

	B	S.E.	Wald	df	Sig.	Exp(B)	95% C.I. for EXP(B) Lower	Upper
Step 1ª Control_Small_Arms(1)	2.751	1.133	5.898	1	.015	15.654	1.700	144.116
Demilitarization(1)	.351	1.099	.102	1	.749	1.421	.165	12.256
Zscore: #Uniformed	.548	.587	.873	1	.350	1.730	.548	5.460
Civil Liberty	1.072	.430	6.213	1	.013	2.920	1.257	6.781
Constant	-6.798	2.474	7.551	1	.006	.001		

a. Variable(s) entered on step 1: Control Small Arms, Demilitarization, Zscore: #Uniformed, Civil Liberty.

The odds of peacekeeper fatalities are 15.654 times greater if the peacekeeping task requires UN personnel to take control of small arms and light weapons (SALWs). The PEMA dataset defines the control of SALWs as their collection, storage, and destruction. The codebook also notes that the task may target specific groups within the conflict (PEMA codebook p. 8). Seizure of SALWs may, therefore, disproportionately impact non-state actors who are likely to have a narrower range of weapons than state actors. This finding is consistent with Dorussen and Gizelis (2013), who found that rebels are most likely to resist policies in which the UN has direct and sole responsibility as it offers no opportunities for gain and impacts their relative bargaining power.

Additionally, the demilitarization task, which required the withdrawal of troops and heavy weapons from specific areas (likely impacting state actors) also proved insignificant in the model. Similar to the findings by Fjelde et al. (2016), the Military Model suggests that a loss of power by the non-state actor, measured by the loss of SALWs, increases the risk of attack against peacekeepers. These results may conflict with Salverda's (2013) finding that attacks against peacekeepers are more likely when the insurgent group is the more powerful belligerent (p. 717). If peacekeepers confiscate SALWs during a conflict, resulting in a shift in power favoring the government, disarmed non-state actors must rely on peacekeepers for protection from attacks by their adversaries, including the state. The Military Model, however, predicts violence against peacekeepers in this scenario, suggesting that the insurgents believe the peacekeepers will fail to protect them from attacks from other belligerents while simultaneously disrupting their weapon supply networks.

The two non-PEMA independent variables in the Military Model were the number of uniformed personnel assigned to the mission and the level of civil liberty in the state where the mission operated, as defined in Freedom in the World data. The number of uniformed personnel proved insignificant in the Military Model but significant when the peacekeepers performed non-military-related tasks, as shown in the previous two models. Increasing civil liberties is associated with an increased likelihood of peacekeeper fatalities. Freedom in the World Dataset includes a rating of 1 to 7, with 1 representing the most freedom and 7 the least. The variable reflects the freedom of expression and belief, associational and organizational rights, rule of law, and personal autonomy and individual rights within each country. The Military Model shows that the likelihood of peacekeeper fatalities increases as freedom increases. Distrust and fear of the peacekeepers in the context of a more open environment could make insurgent attacks easier to carry out. Other ethnic, religious, or political tensions, suppressed in a less free environment, may surface, resulting in more significant risks to the peacekeepers.

Discussion

D ata from the PEMA dataset was used to create three models to predict peacekeeper fatalities, focusing on the missions' tasks. Each model also included two other independent variables selected from the list of significant variables from the univariate regression analysis. The selected variables were chosen to provide insight into peacekeeper fatalities within the context of the mandate. While all three models proved statistically significant, only the Military Model produced a significant mission task: controlling small arms and light weapons. This finding is consistent with the existing literature that claims that power distribution impacts the vulnerability of peacekeepers. Some authors, such as Fjelde et al. (2016), suggest that the loss of power by insurgents increases attacks on peacekeepers, while Salverda (2013) found that the more powerful the insurgents, the more likely they are to attack. The results of this study fall within the former category, as the loss of weapons by non-state actors to the peacekeepers could lead to all sorts of concerns stemming from a sense of defenselessness.

Limitations of the Study

T hree limitations of the study are worthy of discussion. First, the inadequate number of empirical studies on peacekeeper security leaves us with a thin understanding of the subject matter's theoretical underpinnings. More robust models are needed to explore various causes of peacekeep-

er fatalities. The second limitation deals with the study's use of the PEMA dataset. This study used only the modality in which peacekeepers assisted in the task. Two other modalities—monitoring and securing tasks—may fundamentally alter the risks for peacekeepers. Exploration into the risks posed by these different modalities could prove useful. The third limitation of the study is its focus on UN peacekeepers and the exclusion of regional forces. Attack predictors may differ for peacekeepers who are regionally and perhaps culturally closer to the crisis.

Significance of the Results

V iolence against peacekeepers is a neglected field of study. The failure to study peacekeeper fatalities is surprising since much research exists on peacekeeping effectiveness, which peacekeeper fatalities undoubtedly impact. Peacekeeping missions occur in violent environments, but knowing the conditions under which attacks against peacekeepers occur would allow decision-makers to reallocate personnel, equipment, or intelligence assets to areas with lower risk of fatalities. Specifically, knowing that insurgents who relinquish their small arms are particularly likely to attack peacekeepers suggests that this scenario needs further exploration to determine whether it is a sense of defenseless that provokes attack or a reaction to the loss of power or even a fear of retribution by the other belligerents.

Kathryn Lambert holds a PhD in Political Science. Her research interest focuses on political violence by non-state actors, exploring concepts such as political predictors of attacks and drivers of target selection. She hopes that her research will contribute to our understanding of violence against those who seek to improve our world, such as international aid workers and UN peacekeepers. She welcomes opportunities for continued research and collaboration.

References

Abbs, L., and Duursma, A. (2024). Tracing the footsteps of peace: Examining the locations of UN peacekeeping patrols. *International Interactions, 50*(6), 975–1004. https://doi.org/10.1080/03050629.2024.2413566

Aid Worker Security Database, retrieved August 20, 2023, from https://www.aid workersecurity.org/

Bangura, M.A. (2010). Prosecuting the crime of attack on peacekeepers: A prosecutor's challenge. *Leiden Journal of International Law, 23*(1), 165–181. https://doi.org/10.1017/S0922156509990379

Bara, C. (2020). Shifting targets: The effect of peacekeeping on postwar violence. *European Journal of International Relations, 26*(4), 979–1003. https://doi.org/10.1177/1354066120902503B

Bloom, E. T. (1995). Protecting peacekeepers: The Convention on the Safety of United Nations and Associated Personnel. *The American Journal of International Law, 89*(3), 621–631. https://doi.org/10.2307/2204182

Di Salvatore, J. (2019). Peacekeepers against criminal violence—Unintended effects of peacekeeping operations? *American Journal of Political Science, 63*(4), 840–858. https://doi.org/10.1111/ajps.12451

Di Salvatore, J., Lundgren, M., Oksamytna, K., & Smidt, H. M. (2022). Introducing the Peacekeeping Mandates (PEMA) Dataset. *The Journal of Conflict Resolution, 66*(4-5), 924–951. [Dataset and codebook]. https://doi.org/10.1177/0022002721 1068897.

Dorussen, H., and Gizelis, T.-I. (2013). Into the lion's den: Local responses to UN peacekeeping. *Journal of Peace Research, 50*(6), 691–706. https://doi.org/10.1177/0022343313484953

Duursma, A. (2019). Obstruction and intimidation of peacekeepers: How armed actors undermine civilian protection efforts. *Journal of Peace Research, 56*(2), 234–248. https://doi.org/10.1177/0022343318800522

Duursma, A., Lindberg Bromley, S., & Gorur, A. (2024). The impact of host-state consent on the protection of civilians in UN peacekeeping. *Civil Wars, 26*(1), 16–40. https://doi.org/10.1080/13698249.2023.2196185

Freedom House. (2023). *Country and Territory Ratings and Statuses, 1973–2023* [Data set]. https://freedomhouse.org/report/freedom-world

Fjelde, H., Hultman, L., & Bromley, S. L. (2016). Offsetting losses: Bargaining power and rebel attacks on peacekeepers. *International Studies Quarterly, 60*(4), 611–623. https://doi.org/10.1093/isq/sqw017

Fjelde, H., Hultman, L., & Nilsson, D. (2019). Protection through presence: UN peacekeeping and the costs of targeting civilians. *International Organization, 73*(1), 103–131. https://doi.org/10.1017/S0020818318000346

Gadler, A. (2010). The protection of peacekeepers and international criminal law: Legal challenges and broader protection. *German Law Journal, 11*(6), 585–608. https://doi.org/10.1017/S2071832200018745

Global Terrorism Database, retrieved August 20, 2023, from https://www.start.umd.edu/gtd

Greenwood, C. (1996). Protection of peacekeepers: The legal regime. *Duke Journal of Comparative & International Law, 7*(1), 185–207.

Haass, F., and Ansorg, N. (2018). Better peacekeepers, better protection? Troop quality of United Nations peace operations and violence against civilians. *Journal of Peace Research, 55*(6), 742–758. https://doi.org/10.1177/0022343318785419

Hoelscher, K., Miklian, J., & Nygård, H. V. (2017). Conflict, peacekeeping and humanitarian security: Understanding violent attacks against aid workers, *International Peacekeeping, 24*(4), 538–565. https://doi.org/10.1080/13533312.2017.1321958

Kaufmann, D., and Kraay, A. (2023). Worldwide Governance Indicators, 2023 Update), retrieved October 19, 2023, from www.govindicators.org

Kim, W., Sandler, T., & Shimizu, H. (2020). A multi-transition approach to evaluating peacekeeping effectiveness. *Kyklos (Basel), 73*(4), 543–567. https://doi.org/10.1111/kykl.12250

Kisangani, E. F., and Mitchell, D. F. (2021). The impact of integrated UN missions on humanitarian NGO security: A quantitative analysis. *Global governance: A review of multilateralism and international organizations, 27*(2), 202–225. https://doi.org/10.1163/19426720-02702005

Kayinamura, R. (2022). Deputy Permanent Representative of the Republic of Rwanda to the United Nations Plenary Session of Special Committee on Peacekeeping Operations on 15 February 2022.

Laerd Statistics. (2017). Binomial Logistic Regression using SPSS Statistics. *Statistical Tutorials and Software Guides.* https://statistics.laerd.com/

Levin, A. (2023). The composition of UN peacekeeping operations and aid worker security. *Journal of Peace Research, 0*(0). https://doi.org/10.1177/00223433231159186

Lundgren, M., Oksamytna, K., & Bove, V. (2022). Politics or performance? Leadership accountability in UN peacekeeping. *The Journal of Conflict Resolution, 66*(1), 32–60. https://doi.org/10.1177/00220027211028989

Mitchell, D. F. (2015). Blurred lines? Provincial reconstruction teams and NGO insecurity in Afghanistan, 2010–2011. *Stability: International Journal of Security & Development, 4* (1), 1–18. http://dx.doi.org/10.5334/sta.ev

Ranganathan, P., Pramesh, C. S., & Aggarwal, R. (2017). Common pitfalls in statistical analysis: Logistic regression. *Perspectives in clinical research, 8*(3), 148–151. https://doi.org/10.4103/picr.PICR_87_17.

Ruggeri, A., Dorussen, H., & Gizelis, T.I. (2017). Winning the peace locally: UN peacekeeping and local conflict. *International Organization, 71*(1), 163–185. https://doi.org/10.1017/S0020818316000333

Salverda, N. (2013). Blue helmets as targets: A quantitative analysis of rebel violence against peacekeepers, 1989–2003. *Journal of Peace Research, 50*(6), 707–720. https://doi.org/10.1177/0022343313498764

Sauter, M. (2022). A shrinking humanitarian space: Peacekeeping stabilization projects and violence in Mali. *International Peacekeeping, 29*(4), 624–649. https://doi.org/10.1080/13533312.2022.2089875

Seet B., and Burnham G.M. (2000). Fatality trends in United Nations peacekeeping operations, 1948–1998. *JAMA, 284*(5):598–603. doi:10.1001/jama.284.5.598

United Nations. (2023, May 19). *Calling peacekeepers 'beating heart' of United Na-*

tions Commitment to more peaceful world, Secretary-General urges continued sup-port for blue helmets [Press release]. https://press.un.org/en/2023/sgsm21803.doc.htm

United Nations. *Global Peacekeeping Data*, retrieved October 27, 2023, from https://peacekeeping.un.org/en/data

United Nations. *United Nations Peacekeeping Fatalities*, retrieved May 14, 2024, from https://peacekeeping.un.org/en/fatalities

United Nations. *United Nations Peacekeeping Missions*, retrieved August 24, 2023, from https://peacekeeping.un.org/en

United Nations. (2008). *United Nations peacekeeping operations: Principles and guidelines.* https://peacekeeping.un.org/sites/default/files/capstone_eng_0.pdf

United Nations. (2000). *Report of the Panel on United Nations Peace Operations* (A/55/305-S/2000/809). https://peacekeeping.un.org/sites/default/files/a_55_305_e_brahimi_report.pdf

United Nations. (n.d.). *Service and Sacrifice.* Retrieved April 2, 2024, from https://peacekeeping.un.org/en/service-and-sacrifice

United Nations. (2017). *United Nations Security Management System, Security Policy Manual.* https://www.un.org/en/pdfs/undss-unsms_policy_ebook.pdf

United Nations Secretary General. (2022). Safety and security of humanitarian personnel and protection of United Nations personnel: Report of the Secretary General (A/77/362). https://documents.un.org/doc/undoc/gen/n22/599/06/pdf/n2259906.pdf?token=N7bVRX28hZJb2LB8iG&fe=true

United Nations Security Council. (1948). Resolution 50 adopted by the Security Council at its 310th meeting, of 29 May 1948. S/RES/50(1948).

APPENDIX A

Peacekeeping Missions

MISSION NAME	REGIONAL LOCATION	DATES OF OPERATION
UN Operation in Burundi (ONUB)	Africa	2004–2006
UN Mission in Côte d'Ivoire (MINUCI)	Africa	2003–2004
UN Mission for Justice Support in Haiti (MINUJUSTH)	Americas	2017–2019
UN Mission in the Central African Republic (MINURCA)	Africa	1998–2000
UN Mission in the Central African Republic and Chad (MINURCAT)	Africa	2007-2010
UN Mission for the Referendum in Western Sahara (MINURSO)	Africa	1991–present
UN Multinational Integrated Stabilization Mission in the Central African Republic (MINUSCA)	Africa	2014–present
UN Multidimensional Integrated Stabilization Mission in Mali (MINUSMA)	Africa	2013–present
UN Stabilization Mission in Haiti (MINUSTAH)	Americas	2004–2017
UN Civilian Police Mission in Haiti (MIPONUH)	Americas	1997–2000
UN Observer Mission in Angola (MONUA)	Africa	1997–1999
UN Organization Mission in the Democratic Republic of the Congo (MONUC)	Africa	1999–2010
UN Organization Stabilization Mission in the Democratic Republic of the Congo (MONUSCO)	Africa	2010–present
UN Mission in Sierra Leone (UNAMSIL)	Africa	1999–2005
African Union-UN Hybrid Operation in Darfur (UNAMID)	Africa	2007–2020
UN Angola Verification Mission III (UNAVEM III)	Africa	1995–1997
UN Civilian Police Support Group (UNCSPG)	Europe	1998–1998
UN Interim Security Force for Abyei (UNISFA)	Africa	2011–present
UN Mission in Ethiopia and Eritrea (UNMEE)	Africa	2000–2008
UN Mission in Bosnia and Herzegovina (UNMIBH)	Europe	1995–2002
UN Mission in Liberia (UNMIL)	Africa	2003–2018
UN Mission in the Sudan (UNMIS)	Africa	2005–2011
UN Mission of Support in East Timor (UNMISET)	Asia & Pacific	2002–2005
UN Mission in the Republic of South Sudan (UNMISS)	Africa	2011–present

UN Integrated Mission in Timor-Leste (UNMIT)	Asia & Pacific	2006–2012
UN Mission of Observers in Prevlaka (UNMOP)	Europe	1996–2002
UN Mission of Observers in Tajikistan (UNMOT)	Asia & Pacific	1994-2000
UN Operations in Côte d'Ivoire (UNOCI)	Africa	2004–2017
UN Observer Mission in Georgia (UNOMIG)	Europe	1993–2009
UN Observer Mission in Liberia (UNOMIL)	Africa	1993–1997
UN Observer Mission in Sierra Leone (UNOMSIL)	Africa	1998–1999
UN Preventive Deployment Force (UNPREDEP)	Europe	1995–1999
UN Support Mission in Haiti (UNSMIH)	Americas	1996–1997
UN Transitional Administration for Eastern Slavonia, Baranja and Western Sirmium (UNTAES)	Europe	1996–1998
UN Transitional Administration in East Timor (UNTAET)	Asia & Pacific	1999–2002
UN Transition Mission in Haiti (UNSMIH)	Americas	1997–1997
UN Operation in Mozambique (UNUMOZ)	Africa	1992–1994
UN Mission in Haiti (UNTMIH)	Americas	1993–1996
UN Confidence Restoration Operation in Croatia (UNCRO)	Europe	1995–1996
UN Transitional Authority in Cambodia (UNTAC)	Asia & Pacific	1992–1993
UN Advance Mission in Cambodia (UNAMIC)	Asia & Pacific	1991–1992
UN Protection Force (UNPROFOR)	Europe	1992–1995
UN Operation in Somalia II (UNOSOM II)	Africa	1993–1995
UN Operation in Somalia I (UNOSOM I)	Africa	1992–1993
UN Assistance Mission for Rwanda (UNAMIR)	Africa	1993–1996
UN Observer Mission Uganda-Rwanda (UNOMUR)	Africa	1993–1994
UN Angola Verification Mission II (UNAVEM II)	Africa	1991–1995

APPENDIX B

Independent Variables used in the Exploratory Analysis

Independent Variable	Source	Univariate sign.
Disarmament & Demobilization	PEMA	.502
Democratization	PEMA	.217
Reintegration	PEMA	.103
Control of Small Arms & Light Weapons	PEMA	.001
Demilitarization	PEMA	.058
Arms Embargo	PEMA	.309
Human Rights	PEMA	.122
Children's Rights	PEMA	.056
Sexual and Gender-based Violence	PEMA	.112
Use of Force	PEMA	.903
Police Reform	PEMA	.087
Military Reform	PEMA	.309
Justice Sector Reform	PEMA	.177
Transitional Justice	PEMA	.694
Corrections Reform	PEMA	.112
Border Control	PEMA	.593
Demining	PEMA	.459
State Authority Extension	PEMA	.035
Electoral Security	PEMA	.943
Electoral Assistance	PEMA	.970
Civil Society	PEMA	.116
Public Information	PEMA	.362
National Reconciliation	PEMA	.978
Local Reconciliation	PEMA	.058
Regional Reconciliation	PEMA	.217
Humanitarian Relief	PEMA	.450
Economic Development	PEMA	.536
Refugees/Internally Displaced Persons	PEMA	.141
Gender	PEMA	.362
Legal Reform	PEMA	.751
Political Party Assistance	PEMA	1.000
Voters Education	PEMA	.740
Ceasefire	PEMA	.943

Public Health	PEMA	1.000
Media Development	PEMA	.754
Civilian Protection	PEMA	No data
Offensive Operations	PEMA	No data
Peace Process	PEMA	No data
Resources	PEMA	No data
Cultural Heritage Protection	PEMA	No data
Power Sharing	PEMA	No data
Human Development Index	United Nations	.596
Political Rights	Freedom House	.004
Civil Liberties	Freedom House	.003
Country Freedom Status*	Freedom House	.003
Voice and Accountability**	WGI	.025
Stability**	WGI	.009
Effectiveness**	WGI	.061
Number of Troops	United Nations	.006
Number of Uniformed Personnel	United Nations	.005
Number of Police	United Nations	.441
Top Contributing Country from same Region as Mission	United Nations	.110
Military Expenditure as % of GDP for TCC	SIPRI Military Expenditures	.906
Region of Mission	United Nations	.451
Aid Victim Attacks	AWSD (Nexus Uni for pre 1997 event years)	.003
Frequency of Terrorism	GTD	.033

Bold: Denotes significant variable in the univariate regression analysis

*Free and party free categories were merged due to low cell count for the free category

**Values for 1996 used for those event years prior to 1996. The average taken if mission covered more than one state or two partial years.

Security and Intelligence • *Volume 10, Number 1* • *Spring 2025*

The Future of FVEY: Could France Replace New Zealand as an Intelligence-Sharing Partner?

Victoria J. Sengelman[1]

ABSTRACT

This paper discusses international intelligence-sharing partnerships as an essential element in the fight against terrorism and global security. It is imperative that FVEY Community share resources and intelligence to succeed in maintaining a global intelligence picture. Because of intelligence's secretive, sharing does not come naturally to intelligence services, especially with foreign partners. Nevertheless, foreign partners of the United States must be capable and proficient in intelligence collection and sharing. A reexamination of international intelligence sharing relationships, however, is necessary, especially with New Zealand as a current yet sporadic intelligence-sharing partner. As an alternative, this paper explores a potential sharing partnership with France. This paper examines the challenges and capabilities associated with intelligence sharing between these two countries. Additionally, it considers the potential for establishing a more formalized relationship with France.

Keywords: FVEY; intelligence-sharing; Strategic Partnerships; France; New Zealand

El futuro de FVEY: ¿Podría Francia reemplazar a Nueva Zelanda como socio de intercambio de inteligencia?

RESUMEN

Este documento analiza las alianzas internacionales para el intercambio de inteligencia como elemento esencial en la lucha contra el terrorismo y la seguridad global. Es imperativo que la Comunidad FVEY comparta recursos e inteligencia para mantener un panorama global de inteligencia. Debido al secretismo de la inteligencia, compartirla no es algo natural para los servicios de inteligencia, especialmente con socios extranjeros. No obstante, los socios extranjeros de Estados Unidos deben ser capaces y competentes en la recopilación e intercambio de inteligencia. Sin embargo, es necesario reexaminar

1   OSINewMath@hotmail.com

                    doi: 10.18278/si.10.1.6

las relaciones internacionales de intercambio de inteligencia, especialmente con Nueva Zelanda como socio actual, aunque esporádico. Como alternativa, este documento explora una posible alianza con Francia. Este documento examina los desafíos y las capacidades asociadas con el intercambio de inteligencia entre ambos países. Además, considera la posibilidad de establecer una relación más formalizada con Francia.

Palabras clave: FVEY; intercambio de inteligencia; asociaciones estratégicas; Francia; Nueva Zelanda

五眼联盟的未来：法国能否取代新西兰成为情报共享伙伴？

摘要

本文探讨了国际情报共享伙伴关系作为"打击恐怖主义和维护全球安全"一事的重要组成部分。五眼联盟必须共享资源和情报，才能成功维护全球情报图景。由于情报的保密性，情报机构并非自然而然地进行情报共享，尤其是与外国合作伙伴共享情报。尽管如此，美国的外国合作伙伴必须具备熟练的情报收集和共享能力。然而，重新审视国际情报共享关系是必要的，尤其是在新西兰作为不定期的当前情报共享伙伴的情况下。作为一种替代方案，本文探究了与法国建立潜在的共享伙伴关系。本文分析了美法两国情报共享一事面临的挑战和能力。此外，本文还探讨了与法国建立更正式关系的可能性。

关键词：五眼联盟，情报共享，战略伙伴关系，法国，新西兰

Introduction

In 1940, as war loomed globally, President Franklin D. Roosevelt, approved a request for U.S. intelligence services to exchange secret technical information with Great Britain (Panda 2020). The first formal agreement, the Atlantic Charter, established strategic goals and deepened cooperation between the countries (Tossini, 2021). In May 1943, both sides signed the Anglo-American Treaty, an agreement to document intelligence sharing. Both countries created a formal division of Axis ground and air operations for the remainder of World War II (Tossini, 2021), where each assumed the intelligence lead for key areas of focus.

In 1946, with U.S. relations facing the ideological and expansionist threat from the Soviet Union, the U.S. and Great Britain revised the 1943 treaty to include Canada, Australia, and New Zealand (Panda, 2020). Shortly after, the Defense Intelligence Agency (DIA) expanded with the Carrol Agreement between the DIA and the British Ministry of Defense, allowing wider exchange between all members of the "Five Eyes" Intelligence Community/ Alliance (FVEY) (Panda, 2020). However, this plan contrasted with the Soviet approach, which tended to limit cooperation and intelligence sharing with Warsaw Pact partners (Hanson, 2019). However, the cooperation gap between FVEY and the Warsaw Pact partners affected the Soviet Union, and FVEY's ability to easily adapt to changing technologies and organizational and bureaucratic changes proved to be of great benefit during the Cold War (Panda, 2020).

This paper discusses the history of intelligence collection, especially signals intelligence (SIGINT), in New Zealand and France and how they contribute to the FVEY partnership. We will proceed to examine the intelligence community of New Zealand and review its historical contributions to intelligence efforts. Separately, New Zealand had withdrawn its participation due to opposition over nuclear ownership differences and a conflict with France that sank one of its Greenpeace ships. This discussion will focus on the history of intelligence in France, highlighting its significant contributions to the Allied partnership during various conflicts. Additionally, we will explore how France approached intelligence sharing with the United States with a degree of caution. The purpose is to discuss exploring France as a replacement for New Zealand as the "Fifth Eye" in the partnership. France has contributed to intelligence collection and sharing for decades and owns SIGINT satellites over Europe, which have contributed tremendously to their neighboring countries. We will discuss how New Zealand continues to take from FVEY while France continues to give, and how moving forward with an FVEY replacement can ensure enhanced SIGINT capabilities and sharing.

Since its formation after World War II, the FVEY group has never had more than five English-speaking members (Panda, 2020). However, the expansion to share with other countries is evolving. In general, the high degree of interdependence during the Second World War between these countries allowed the sharing of information, including raw and unprocessed information. Furthermore, FVEY originated from the SIGINT collaboration between Great Britain and the US during World War II. Canada established a Special Identification Service in 1946, after years of leadership by Britain and the United States (Panda, 2020).

New Zealand as the Fifth Eye

Today, FVEY is the longest formal intelligence association in history, a collaboration between the United States, the United Kingdom, Canada, Australia, and New Zealand

(Tossini, 2021). For more than 70 years, FVEY has proven to be an essential tool to protect the security interests of each country. With the creation of an Appointee Executive of Government Integration inside DIA, the FVEY community works to stay ahead of global dangers and threats (Tossini, 2021). *The details as to how New Zealand becomes an intelligence-sharing partner are limited,* but the arrangement itself does not do a good job of explaining or promoting the country's contributions and benefits (Battersby, 2023).

FVEY publications only mention New Zealand in passing and, apart from tensions over New Zealand's anti-nuclear foreign policy stance in 1985, offer little detail about the country's involvement (Battersby, 2023). In addition, New Zealand's primary motivation for applying to FVEY was originally due to its immersion in the Western Alliance during the Second World War and its intense perception of security vulnerability after the end of the war. Moreover, this was due to seeking comfort from powerful allies and isolating New Zealand from integrated intelligence systems (Battersby, 2023).

New Zealand is the least intelligence contributor in FVEY. However, fears of an external aggressor have diminished over time as FVEY changed and adapted to changing circumstances in terms of New Zealand's contributions and benefits (Battersby, 2023). Moreover, this transformation was a self-fulfilling prophecy, and New Zealand's role evolved largely because of its involvement in the arrangement from

the start. However, FVEY became increasingly useful to New Zealand as geopolitical tensions rose in the South Pacific (Battersby 2023). Moreover, New Zealand's value to the agreement grew accordingly as communications technology advanced and US political priorities changed (Battersby, 2023). Looking to the future, as its past suggests, New Zealand may need to increase its intelligence sharing to maintain the established mutual benefits that participation in FVEY currently offers (Battersby, 2023).

New Zealand's Intelligence Community

The Intelligence and Security Committee (ISC) oversees the Government Communications Security Bureau (GCSB) and New Zealand Security and Intelligence Service (NZSIS), reviewing policies, management, and spending. GCSB follows policies set by the New Zealand government (Rogers, 2015). The ISC validates certain actions and reviews annual reports, bills, and petitions (Rogers, 2015). In 2012, GCSB's involvement in Kim Dotcom's case revealed confusion within the intelligence community. New Zealand's government arrested Kim Dotcom, who ran a website sharing media under U.S. copyright laws. GCSB's involvement in his extradition case shows a lack of communication within the intelligence community about national security and highlights the problems with its intelligence sharing system (Rogers, 2015).

Investigations revealed a lack of high-level intelligence policy, limited understanding of future operations, unclear operational priorities, and a culture accepting low performance (Rogers, 2015). However, the New Zealand intelligence community needed reformation, but accountability was not transparent (Rogers, 2015). Even more so than that, tied to its economic and foreign policies, New Zealand aims to maintain sovereignty and advance security globally (Rogers, 2015). Additionally, the New Zealand intelligence community focuses on security challenges from state-based and non-traditional actors. The extradition of Kim Dotcom provided a platform for Edward Snowden, Julian Assange, and Glenn Greenwald to inform New Zealanders about GCSB's capabilities in collecting intelligence on New Zealand residents (Rogers, 2015).

New Zealand's intelligence community comprises two organizations. The GCSB gathers foreign intelligence by intercepting communications, cooperating with other entities, collecting information legally, and decoding foreign communications (Rogers, 2015). The NZSIS acquires, correlates, and evaluates human intelligence (HUMINT) for national security purposes. Neither agency provides high-level policy advice (Rogers, 2015). The lack of clarity over national security and the broadening of GCSB's objectives have made it difficult to develop a high-level policy for the intelligence community. Moreover, this has significant implications as officials can share obtrusive information-gathering capabilities without a policy (Rogers, 2015).

New Zealand's Role in FVEY

Furthermore, FVEY shares SIGINT, including raw data, tradecraft, and analysis. FVEY community members can engage in operations together, ranging from simple analysis to covert operations against other nations (Ball, 2023). Although the primary focus is on signals intelligence, they share intelligence from various sources. Hence, FVEY stands out among other intelligence-sharing arrangements due to its longevity, resilience, and unity of worldview among its members (Ball, 2023).

Besides, the Fifth Eye, New Zealand, is of great interest to those who study FVEY intelligence, as it is the smallest and most resource-constrained member of this partnership (Battersby, 2023). Despite this, New Zealand is an active participant, and its role in this arrangement has attracted attention (Battersby, 2023). The details of the union of New Zealand with the alliance are not well-documented. Therefore, its contribution and benefits are vaguely known. As a result, New Zealand is known as the "Ghost Eye" (Battersby, 2023).

New Zealand needs to increase its contributions in intelligence sharing. Although New Zealand has lagged behind other FVEY members, the alliance has adapted to include New Zealand's contributions (Battersby, 2023). However, to maintain the mutual benefit of the alliance, New Zealand may need to expand its intelligence capability (Ball, 2023). During the Second World War, New Zealand played an active role in the intelligence operations of both Britain and Australia. This involvement included the interception of Japanese Naval communications and the provision of timely warnings to Australia of the presence of Japanese submarines off Sydney Harbor in 1942 (Chalk, 2020).

Consequently, New Zealand emerged as an initial UKUSA connection as the arrangement began to take shape. Even so, due to the absence of an appropriate intelligence apparatus, New Zealand had to create a military SIGINT unit to prove its alliance-worthiness to UKUSA (Chalk, 2020). Although New Zealand worked diligently to meet the minimum requirements for joining FVEY, it did not establish a foreign intelligence service like MI6 or ASIS at that time. Even in 2023, New Zealand has yet to set up a foreign intelligence service (Ball, 2023).

At first, New Zealand was not a part of the original UKUSA agreement. However, the British hoped to eventually include New Zealand, considering their significant contribution to SIGINT during the war (Battersby, 2023). The United Kingdom welcomes the sharing of intelligence work, in which New Zealand also plays a role. However, the post-war Australian Defense Signals Bureau (DSB) was a joint initiative between the UK, Australia, and New Zealand (Battersby, 2023). Furthermore, this suggested that New Zealand's inclusion in UKUSA was not merely a courteous gesture but a recognition of New Zealand's long-standing integration into the intelligence services of its main wartime allies. Even so, New Zealand needed to establish an intelligence service before formally joining the UK/U.S. in 1956 (Ball, 2023).

Moreover, New Zealand played a pivotal role in FVEY, particularly in the Southeast Asia and Southwest Pacific regions. The Tangimoana Central North Island SIGINT listening station and the Waihopai listening station were critical components of the Echelon spy network, intercepting radio and satellite communications from various foreign nations, including China, Vietnam, Japan, Egypt, and East Germany (Battersby, 2023). First, the country's location made it an ideal monitoring station for the countries of interest in the Pacific, such as China and Japan (Battersby, 2023). Second, despite claims that New Zealand monitored terrorism nearby, there was almost complete absenteeism of terrorism in the Southern Pacific. The country's policymakers were also vocal in opposing the innumerable French nuclear tests at Moruroa and Fangataufa atolls between 1974 and 1996 (Battersby, 2023). However, while Southeast Asia was closer to Australia, it remained unclear if New Zealand had any specific role in intelligence in that region at first (Battersby, 2023).

New Zealand's participation in FVEY in the past has been invaluable. The GCSB and SIGINT stations at Tangimoana and Waihopai contributed to FVEY's success. Waihopai was even funded by the National Security Agency (NSA), and New Zealand's government facilities (Battersby, 2023). Furthermore, New Zealand's involvement has been essential to the alliance's continued success, especially with the development of new satellite communications technology (Ball, 2023). Additionally, Tangimoana intercepts radio communications in the Pacific to monitor New Zealand's maritime Exclusive Economic Zone. Waihopai is part of the FVEY interception system, which uses cyber-based technologies for electronic interceptions. Furthermore, Waihopai's usefulness in FVEY remains uncertain. In 1984, New Zealand banned nuclear-armed and powered vessels from its ports, resulting in the withdrawal of the U.S. defense guarantee under the ANZUS Treaty. However, Waihopai was built to make New Zealand less reliant on intelligence from other nations. New Zealand's expulsion from the UKUSA in 2006 was due to the nuclear ships ban (Ball, 2023).

How Did New Zealand's Intelligence Landscape Evolve in Response to Changing Geopolitical Tensions?

The NSA defended New Zealand's involvement in an intelligence agreement even though some Washington officials wanted to remove it (Battersby, 2023). Additionally, FVEY can withstand periods of reduced cooperation between member countries as New Zealand entered one of those periods (Battersby, 2023). As a result, the United States reduced the intelligence it shared with New Zealand (Ball, 2023). In July 1985, the French Direction Générale de la Sécurité Extérieuré (DGSE) bombed Greenpeace's Rainbow Warrior in Auckland harbor (Ball, 2023). Furthermore, the incident raised concerns that FVEY failed to detect the attack (Ball, 2023). Even more so than that, some New Zealand intelligence officials suggested

that the United States had knowledge of the plot but chose not to share it due to New Zealand's anti-nuclear stance. This issue caused speculation that there was a severe consequence for New Zealand's "errant behavior" (Battersby, 2023). However, the issue of nuclear ships strained the UKUSA relationship, and New Zealand retaliated by deciding to stop sharing intelligence collected by its two radio outposts (Ball, 2023).

In the mid-1970s, New Zealand continued to receive intelligence information from the other members of FVEY for the next two decades (Ball, 2023). However, during the 1987 Fiji coup, New Zealand did not share any intelligence about the coup or its aftermath, which caused a major rift in FVEY. New Zealand has since stopped sharing intelligence, forcing other members to fill the gap in Southeast Asia (Ball, 2023). Currently, New Zealand collects and receives SIGINT from other sources but rarely obtains HUMINT or military operations information (Ball, 2023).

In 1984, new Prime Minister David Lange became concerned about nuclear weapons and French nuclear testing (Ball, 2023). New Zealand's membership in the alliance had become more valuable, offering a permanent connection to a global intelligence network that it could not achieve alone (Ball, 2023). Furthermore, New Zealand prioritized intelligence on French activity in the South Pacific, with FVEY supporting its investigations from 1985 onwards (Hanson, 2019). However, New Zealand's ban on nuclear ships and reduced U.S. intelligence through FVEY may have impacted its ability to stop French nuclear tests in the region (Ball, 2023).

Although U.S. administrations have officially withheld intelligence due to the anti-nuclear issue, FVEY agencies continued to work together (Ball, 2023). However, in 2006, media in the National Archives revealed the extent of the GCSB's electronic interception operations against countries such as Fiji, Vietnam, Laos, South Africa, and Argentina (Ball, 2023). However, the GCSB depended on its British and U.S. allies to collect French communications that were out of range of its monitoring stations. Following the Rainbow Warrior incident, the GCSB requested the Government Communications Headquarters (GCHQ) and the NSA to monitor specific targets in France, which they did (Ball ,2023).

In 2006, New Zealand fully reintegrated into the FVEY intelligence-sharing network, which recognized the alignment of intelligence interests between the US and New Zealand. Moreover, New Zealand's expanded access to FVEY systems and technology allowed it to intercept communications in various parts of the world (Ball 2023). Although the Snowden leaks revealed the country's vast intelligence collection capabilities, most criticisms focused on its eavesdropping activities on friendly South Pacific Island nations (Ball, 2023). Even more, recent allegations of New Zealand's involvement in mass surveillance on behalf of the NSA and connections with torture and rendition programs have also raised questions.

However, the lack of historical intelligence material accessibility makes it difficult to investigate New Zealand's intelligence activity 50 years ago, so it is unlikely that current inquiries will be successful (Ball, 2023).

New Zealand was excluded from FVEY until 2006 after a ban on nuclear ships. Although from 1985 to 2006, New Zealand was able to find niche areas to maintain its FVEY connections, its policies for SIGINT collection were lacking (Battersby, 2023). New Zealand was criticized for its general lack of intelligence in integrating SIGINT into a national strategy, and for obtaining much intelligence information that was futile to them. Despite this, the benefits of FVEY membership are still significant when compared to the information provided. New Zealand's anti-nuclear policy resulted in the loss of ANZUS benefits, however, FVEY was too important to lose to New Zealand (Battersby, 2003). The NSA defended New Zealand's participation in the intelligence arrangement and maintained its resilience, even during periods of reduced cooperation between member countries (Battersby, 2003).

According to a review of Intelligence and Security in 2016, New Zealand receives one hundred times more intelligence than it contributes to FVEY (Cooper, 2018). Still, it shows New Zealand must increase its intelligence collection efforts and provide as much intelligence as it receives within FVEY (Battersby, 2023). While New Zealand will continue to benefit from the intelligence it receives from FVEY, it can no longer rely solely on the alliance for foreign intelligence (Battersby, 2023). If other countries are conducting intelligence operations in the South Pacific, New Zealand must enhance its ability to detect and respond to those activities to serve its vital interests (Ball, 2023).

Additionally, the nuclear ship issue has put pressure on UK-U.S. relations. Furthermore, New Zealand decided to retaliate by ceasing all sharing of collected intelligence by its two radio outposts after the U.S. suspended defense guarantees under the Australia-New Zealand-U.S. security deal (Pfluke, 2019). Over the next two decades, New Zealand continued to receive intelligence collected by FVEY but refused to share its intelligence. For the next two decades, New Zealand's absence led to the FVEY losing valuable intelligence in Southeast Asia and having to fill the gaps (Pfluke, 2019).

Currently, New Zealand retains the SIGINT material it collects itself but also receives other SIGINT materials (Ball, 2023). However, it has received few U.S. HUMINT products or assessments. It also receives no information about military operations except in rare cases where New Zealand is participating, such as in Afghanistan (Ball, 2023). Despite these arrangements, New Zealand maintains its alliance with Australia and station liaison officers in various Australian intelligence agencies (Ball, 2023). As a result, New Zealand and U.S. liaison officers must be housed separately in Australian agencies and given separate briefings, and there are strict limitations on what Australia can share with New Zealand. This event has

resulted in a new category of "four eyes only" material, which excludes New Zealand (Ball, 2023).

Contradiction Between Alliance and Agreement in FVEY

New Zealand's role evolved, reflecting a changing world. The FVEY is not a coalition between countries with different national interests, but rather a pact between intelligence agencies that have established a tradition of working together. In 1965, Australia and New Zealand sent troops to Vietnam, but the UK did not (Ball, 2023). The tension between the U.S. and the UK and the U.S. and Australia in the 1970s sometimes made intelligence sharing difficult. By the time New Zealand adopted its anti-nuclear policy in 1985, FVEY had learned to adapt to changes in the political climate (Ball, 2023). Ultimately, FVEY focused on exchanging information between intelligence agencies, rather than aligning the policy positions of their governments. It is an agreement, not an alliance (Ball, 2023).

When political leaders disagree on policy positions, they may stop formal intelligence-sharing processes (Ball, 2023). This issue happened during the 1985 nuclear impasse, and it is unclear what intelligence was shared separately at that time. New Zealand also received foreign intelligence outside of the FVEY connections, leading to increased intelligence sharing in the mid-1980s. However, New Zealand also withheld intelligence on the Fiji coups in 1987, which may have been a symbolic reaction to the U.S.'s lack of interest in the region (Ball, 2023). Despite this, New Zealand remained a member of FVEY and intelligence sharing continued. The recent U.S. responses to China's growing influence in the South Pacific have highlighted the U.S.'s long-standing indifference to the region. Eventually, the period of difficulty passed, as had happened among members in the past (Ball, 2023).

In New Zealand, the nuclear ship ban was a popular move among its citizens and remains in place today (Ball, 2023). There is no evidence to suggest that the country's leadership ever intended to harm ANZUS, and it was the U.S. who withdrew its defense guarantee, reflecting their lack of interest in the South Pacific (Ball, 2023). Furthermore, the development of satellite communications detached the interception business from geographic realities, leading to more capable SIGINT interception and strengthening New Zealand's value to FVEY (Ball, 2023).

However, the Tangimoana Central North Island SIGINT listening station and Waihopai SIGINT station were critical components during the Cold War by intercepting radio and satellite communications from various foreign nations, including China, Vietnam, Japan, Egypt, and East Germany (Ball, 2023). The country's location made it an ideal monitoring station for countries of interest in the Pacific, such as China and Japan. Despite New Zealand's claim of monitoring terrorist activities in the region, they have stayed entirely absent from the South Pacific region. Regardless of who paid for the Waihopai spy base, the NSA or New Zealand with Australian assistance, this response is unlikely to happen (Ball, 2023).

The Future of New Zealand as the "Fifth Eye"

Ultimately, New Zealand's membership in FVEY was inherited from the Second World War due to a strong perception of its vulnerability in the Indo-Pacific region (Ball ,2023). In 2021, riots broke out in Honiara, the capital of the Solomon Islands, due to the country's increased ties with China. Chinese businesses were targeted, and Australian and New Zealand forces were deployed. Chinese police advisors were also present. After the situation was under control, Honiara signed an agreement with Beijing for future Chinese law and order assistance. A draft agreement hinted at Chinese military aid, which prompted the United States to increase its military commitment to the region through the AUKUS agreement (Battersby, 2023). Political instability in the South Pacific could lead to significant powers getting involved, making intelligence gathering crucial for the FVEY. New Zealand's nuclear ship ban and the reduction of U.S. intelligence through FVEY may have weakened their ability to reduce French nuclear tests in the South Pacific (Battersby, 2023).

France as the Fifth Eye

France's Early Use of SIGINT

Before World War I, the French were leaders in cryptography. They could decrypt diplomatic messages between Germany's Foreign Office in Berlin and its embassy in Paris. The French Army monitored German wireless communication and gained a thorough understanding of the German Army's peacetime communications. This expertise helped the Allies during the early stages of the war when the Germans advanced into France. Although it did not completely halt the German offensive, it helped stop it during the "Miracle on the Marne" and the following "Race to the Sea." The French dominance of cryptography is a testament to their intelligence and ability to utilize technology to their advantage (Canuel, 2013).

Early in the history of the French intelligence service, Bureau Central de Renseignements et d'Action (BCRA), the only SIGINT activity outside of France occurred in the Vichy Empire. This activity was strategically significant as Vichy was planning to defend its colonies by force. The Petainist authorities in French North Africa and West Africa had the highest intelligence-gathering capacity, as they needed to obtain "battlefield intelligence" from SIGINT and HUMINT sources. Vichy's empire faced more external threats than the mainland, including German intervention in North Africa, Italian designs on Tunisia and French Somaliland, Spanish threats to French Morocco, and Anglo-Gaullist incursions into any Vichy colony (Canuel, 2013).

During World War II, Vichy relied on French African armistice forces to monitor and intercept signals from potentially hostile forces. Vichy's intelligence capability strengthened as the threat of incursion rose. Rivet and Ronin's SR organization provided valuable SIGINT to the British, and French intelligence operatives sent 2,748 "Cadix" decrypts to London. The Tunisian campaign allowed French Enigma decrypts to be sent to military command, while the Italian campaign relied on agents' reports. French intelligence played a crucial role in preparing for the Normandy landings, but Rivet failed to keep it out of the political conflict between Gaullists and their rivals (Canuel, 2013).

The victorious Allied advance and the rapid establishment of France's transitional government allowed French intelligence services to devote more resources than expected to analyze signals intelligence (Thomas, 2008). Finally, the remarkable commitment and success of an improvised intelligence service working in extremely adverse conditions inside Vichy France gave the lie to any suggestion that Pétain's regime lacked either the will or the means to continue working against the Axis powers (Thomas, 2008). The experience of Rivet's SR staff and Bertrand's "Cadix" teams confirms that once learned, intelligence skills are neither readily surrendered nor easily curbed. France's intelligence capabilities should therefore caution us against viewing France as a regime lacking the capacity for independent diplomacy and strategic decision-making (Thomas, 2008).

Eventually, Vichy's SIGINT activities were bound up in the contest between the protagonists and opponents of collaboration in Pétain's regime (Thomas, 2008). However, throughout the 1940s, the threat perceptions that informed Vichy's military intelligence gathering remained unaltered from those of 1939–40 (Thomas, 2008). The Axis was the enemy, and the Allies were at least potential friends. The biggest lesson that the recipients of the Vichy signal learned was that those who provided the signal worked for a regime far less monolithic than commonly imagined (Thomas, 2008).

France's SIGINT Contributions

In contrast to New Zealand's more passive role, France has demonstrated initiative-taking and extensive intelligence capabilities. Although France was one of the victors in 1945 and set out to create a world-class signals intelligence agency, Paris did not get the invitation to sign the 1946 UKUSA agreement, and French leaders did not want to join the AUSCANUKUS community (Canuel, 2013). Determining how this occurred is not as simple as it seems. France played a key role in early peacetime SIGINT cooperation, the breaking of the Enigma code, and cooperation with Britain, with whom it continued to cooperate even after France's fall in 1940 (Canuel, 2013). Even so, post-war French ambitions and Anglo-Saxon suspicions prevailed at the dawn of the Cold War, resulting in a unique dynamic between these three powers (Canuel, 2013).

At the start of the Cold War in 1947, France used an extensive diplomatic intelligence network established in the 1930s, in Africa and Asia. The post-war reorganization of the French intelligence services brought together ex-combatants, ex-soldiers, and police officers who worked under the Vichy government for the common cause of defeating communism (Faligot, 2001). In Indochina and Algeria, they actively engaged in subversive wars, while in Eastern Europe, they conducted communication wars (Faligot, 2001). At the same time, they opposed US actions. After the Gulf War, France actively developed an anti-satellite system which played a crucial role during conflicts in Central Europe. French strategists consider it a significant step towards establishing a European communication system (Faligot, 2001).

Despite the difficulties of political and bureaucratic competition, all aspects of SIGINT operations, postwar France managed to fulfill all activities related to the collection and use of domestic and foreign radio traffic (Canuel, 2013). However, the U.S. and Britain may have overlooked this immense potential as they sought to set up new levels of cooperation in areas not seen before in peacetime.

SIGINT Satellite Monopoly

Some imaging satellites contain sensors that collect images in the infrared (IR) part of the electromagnetic spectrum. Infrared light is between the visible and microwave portions of the electromagnetic spectrum and has many wavelengths, just as visible light has wavelengths from red to violet (Crothers, 2009). The SPOT-4 earth observation satellite, owned, and operated by France, has an additional sensor capable of imaging objects in the shortwave infrared (SWIR) range (Crothers, 2009). France is the only EU country that controls the military satellite imaging system (Clough, 2004). In 1986, the first SPOT satellite with a ground-based instrument entered orbit (Florini, 1988). Its multispectral resolution is 20 meters, and its monochromatic resolution is 10 meters. SPOT Image Corporation follows open skies policies but has decision-making autonomy (Florini, 1988).

There are no restrictions on collecting military information. SPOT's capabilities depend on commercial demand. France's SPOT satellite system has Helios, a military version with 1 1-meter resolution and electronic intelligence (Florini, 1988). Other countries utilize SPOT for various purposes, and France is at the forefront of satellite technology with its upcoming Helios launch (Florini, 1988). France has developed two advanced satellite monitoring programs, SPOT and Helios, which could serve as the foundation for a regional or global satellite monitoring agency. Other European countries are already involved in the Helios program and hold small shares of SPOT. First, innovative technologies can make satellite data even more valuable for monitoring purposes, enhancing global security efforts (Florini, 1988). Second, satellite technology for SIGINT collection and Cyber Warfare is becoming the dominant topic in intelligence (Florini, 1988). Acquiring France as an intelligence partner in FVEY benefits the U.S. tremendously with additional infor-

mation from France's imagery satellites (Canuel, 2013).

France, FVEY, and Missed Opportunities

The FVEY agreement emerged from the UKUSA Agreement of 1943, which governed cooperation between the United Kingdom and the U.S. in matters related to SIGINT. Over the years, this agreement expanded to include Canada, Australia, and New Zealand, forming a circle of trusted countries. However, one country remained outside this circle—France. Despite participating in the defeat of the Axis, France did not want to be a part of the intelligence-sharing agreement. Agreements such as the UK-U.S. deal rely on compromises on sovereignty that appear unacceptable to France (Canuel, 2013).

After WWII, the United Nations Security Council granted France a seat and secured an occupation zone in Germany (Canuel, 2013). The English-speaking powers did not trust France, and France did not want to be a part of the intelligence-sharing agreement (Canuel, 2013). There are no contemporaneous sources that suggest the signatories contemplated extending membership to France or that Paris seriously considered approaching London and Washington to be active participants (Canuel, 2013).

However, bilateral relations between French SIGINT agencies and those of the NSA and GCHQ existed, but these exchanges were quite limited in nature and never involved the transparency underlying the UKUSA Agreement (Canuel, 2013). Even though France and her allies did exchange intelligence derived from communication intercepts, the information was limited to what one wanted to divulge and when one chose to do so (Canuel, 2013). Consequently, France remained an outsider to the FVEY, and their stance towards the UKUSA Agreement remained ambivalent (Canuel, 2013). France has SIGINT assets that could have been of great interest to their allies. However, France's potential contribution to the SIGINT community remained unexplored due to a lack of cooperation (Canuel, 2013).

Immediately after the end of World War II, France invested heavily in rebuilding first-class signals intelligence facilities in Paris. They set up intercept stations and supply bases for ships, aircraft, and submarines required to collect intelligence close to the borders and coastline of any given target of interest (Canuel, 2013). However, France's resources did not fill any specific gap in the UKUSA Agreement as her allies had already surpassed French agencies in analysis techniques and computing capabilities (Canuel, 2013).

Paris was neither invited nor willing to join the Anglo-American agreement negotiated by London and Washington in 1946 (Canuel, 2013). This crucial moment in the development of the "special relationship" between the two wartime allies captured the spirit and practice of the SIGINT cooperation that had evolved during the war (Canuel, 2013). France's leaders aspired to regain their country's posi-

tion of prominence on the international scene. This response has led them to pursue independent courses of action within the larger Atlantic partnership while maintaining targeted bilateral relations where appropriate (Canuel, 2013). However, this combination of French desire and Anglo-Saxon suspicion negated the potential for positive synergies in signals intelligence and led to a stark contrast in the dynamics between the three Western victors in the early years of the Cold War (Canuel, 2013).

France has a significant opportunity to create positive harmony in the realm of SIGINT by joining the Anglo-American agreement. With France's local knowledge and vast resources, combined with the existing techniques and capabilities of the UKUSA allies, this partnership could become unbeatable (Canuel, 2013). Furthermore, France's independent course of action may have seemed like the right choice at the time, but joining the Anglo-American agreement would greatly enhance global intelligence coverage. Therefore, it is time for France to make the right decision and become a part of this alliance (Canuel, 2013).

FVEY and France Moving Forward

In 2009, the NSA offered to make the DGSE a member of the exclusive FVEY club (Tréguer, 2017). Apparently, the "Sixth Eye" deal failed over the Central Intelligence Agency's (CIA) refusal to conclude a no-spy agreement with France, and in 2011, a more modest cooperation was eventually signed between the NSA and the DGSE under the form of a memorandum, most likely the LUSTRE agreement revealed in 2013 by NSA whistleblower Edward Snowden (Tréguer, 2017). The "Sixth Eye" agreement could have increased intelligence sharing worldwide with the addition of France, but after the Snowden incident, France lost trust mainly with the U.S. as an intelligence-sharing partner (Tréguer, 2017). Slowly, the U.S. attempted to mend relations with France to improve intelligence sharing.

However, in 2021, the U.S. won the AUKUS deal, a nuclear submarine-sharing partnership between the U.S., the UK, and Australia. France became infuriated, as France was about to complete their deal to develop nuclear submarines for Australia (Panda, 2021). It is important to address the criticisms of the AUKUS agreement by the French. While it is understandable that there are concerns and doubts that the U.S. has not sidelined France in the Indo-Pacific (Panda, 2021). Moreover, it is not just the U.S. that supports this decision; Japan and Taiwan have also positively received it (Panda, 2021). It is also not fair to say that the U.S. has completely excluded the European Union (EU) from having a role in the security of the Indo-Pacific (Panda, 2021). The joint communication on the Indo-Pacific cooperation strategy between the two countries clearly stated that while the EU does not intend to take a leading military role in the region, it still has a responsibility to play an active role in ensuring maritime security, cybersecurity, and counter-proliferation (Panda, 2021).

As for France's strategy, it is worth noting that it is not solely based on an arms contract. Furthermore, the arms contract with Australia is only one element of a broader strategy (Panda, 2021). France's interests in the Indo-Pacific are unique among EU member states, as more than 1.6 million French citizens live in overseas territories in the region, and over 90% of France's exclusive economic zone is in the Indo-Pacific (Panda, 2021). Hence, France is not just a spectator in the region but a resident power with sovereignty interests. However, it is vital to see the bigger picture and understand that other nations have a role to play in the Indo-Pacific region. By working together, we can ensure the safety and security of that region (Panda, 2021).

Conclusion

FVEY membership benefits New Zealand with intelligence access that it could not obtain alone. New Zealand's involvement in World War II has increased its value to the alliance (Pfluke, 2019). New Zealand's geopolitical environment is increasingly unstable, as tensions in the Indo-Pacific impact the country (Ball, 2023). The changing geopolitical landscape in the South Pacific demands that New Zealand increase its intelligence efforts, as its current contribution is no longer sufficient (Battersby, 2023). New Zealand has been a secretive member of FVEY for over 70 years, keeping decades-old records locked away. This lack of transparency has resulted in a lack of public awareness and trust (Battersby, 2023).

France's intelligence apparatus underwent a purge after World War II, which resulted in the country's establishment in the anti-Soviet camp during the Cold War (Canuel, 2013). Despite being allies during World War II, France was not included in the UKUSA Agreement, which was negotiated in 1946 between London and Washington (Canuel, 2013). However, France chose not to participate in this agreement, which left it isolated from the rest of the intelligence services that had been significantly altered by the war. France was no longer considered an intelligence power on an international level due to its lack of involvement in the growth of electronic and scientific intelligence during the war (Canuel, 2013).

In brief, the United States faces new adversaries in great power competitions today as we see growing problems with Russia, China, North Korea, and Iran. As the world becomes increasingly complex, it is essential to have a system in place to safeguard against potential threats. FVEYs serve as the eyes and ears of this system, but it is essential to address the weakest links to ensure its effectiveness. With China expanding its territorial claims in the South Pacific, Australia must take a leading role in gathering intelligence for that region. New Zealand has expressed willingness to receive intelligence rather than contribute, but all members must work together to create a comprehensive intelligence picture (Battersby, 2023). Between their lack of intelligence sharing and their inability to allow nuclear vessels into their ports, New Zealand has proven difficult to work with as the

"Fifth Eye" (Battersby, 2023).

France has proven their incredible contribution as an intelligence-sharing partner throughout the history of cryptography (Canuel, 2013). But a lack of trust with the U.S. has stopped all potential agreements to make France the "Sixth Eye" (Panda, 2020). France continues to be an intelligence-sharing partner with other FVEY nations, and with French colonies in the South Pacific, France can cover intelligence-gathering sites where New Zealand is already lacking (Faligot, 2001). France also contributes to the global intelligence picture with their satellites for imagery and SIGINT collection (Crothers, 2009).

In the aftermath of the 2015 terrorist attacks in Paris, France and the U.S. came together to enhance their intelligence-sharing efforts. This partnership is crucial in combating the threat of terrorism and ensuring the safety of citizens in both countries. As a result, both nations worked together to prevent terrorist attacks and began rebuilding their relationship (Bertuca, 2015). This event was a step towards intelligence cohesiveness between the U.S. and France. However, the U.S. needs to extend an olive branch to France to repair the relationship. The AUKUS deal was a major blow to France, and the U.S. should take heed in correcting their dialogue with France (Panda, 2021). France continues to be a great contributor to intelligence worldwide and would prove to be a beneficial contributor to FVEY. With their satellite monopoly in Europe and SIGINT capabilities, France can contribute more than New Zealand has contributed to FVEY in decades.

Victoria Sengelman holds a BA in Intelligence. Her primary areas of research are signals security and intelligence, signals intelligence analysis, strategic leadership, information warfare, military intelligence, and intelligence collection systems and methods. She welcomes opportunities for continued research and collaboration.

References

Annual Report 2021—New Zealand Security Intelligence Service. https://www.nzsis.govt.nz/assets/NZSIS-Documents/NZSIS-Annual-Reports/NZSIS-Annual-Report-2021.pdf.

Ball, Ris, and Murray Place. (2023). "The New Zealand Intelligence Community and Effects Operations: The Covert Action Dilemma." *Intelligence and National Security* Vol. 39, Iss. 4, 709–728, DOI: 10.1080/02684527.2023.2274142.

Battersby, John, and Rhys Ball. (2023). "The Phantom Eye: New Zealand and the Five Eyes." *Intelligence and National Security* ahead-of-print (ahead-of-print): 1–18. DOI: 10.1080/02684527.2023.2212557.

Bertuca, Tony. (2015). "U.S., FRANCE PREPARE TO SURGE INTELLIGENCE SHARING AFTER PARIS ATTACKS." *Inside the Pentagon* 31 (46) (Nov 19).

Canuel, Hugues. (2013). "French Aspirations and Anglo-Saxon suspicions: France, signals intelligence and the UKUSA agreement at the dawn of the Cold War," *Journal of Intelligence History*, 12:1, 76–92, DOI: 10.1080/16161262.2013.755021

Chalk, Peter. (2020). "Domestic Counter-Terrorist Intelligence Structures in the United Kingdom, France, Canada and Australia." *Studies in Conflict and Terrorism* Vol. 45, Iss. 7, 626–658, May 3, 2020. DOI: 10.1080/1057610X.2019.1680186.

Cooper, Simone. (2018). "An Analysis of New Zealand Intelligence and Security Agency Powers to Intercept Private Communications: Necessary and Proportionate?" *Te Mata Koi: Auckland University Law Review* 24 (2018): 92–120. https://search.informit.org/doi/10.3316/informit.604403393518820.

Faligot, Roger. (2001). "France, SIGINT, and the Cold War." *Intelligence and National Security*. 16. DOI: 10.1080/714002843.

Hanson, Colin. (2019). "Signals intelligence in New Zealand during the Cold War" - WGTN, 2019. https://www.wgtn.ac.nz/__data/assets/pdf_file/0003/1726923/2019-1-Signals-Intelligence-in-New-Zealand-during-the-Cold-War-David-Filer-2019.pdf.

Panda, Ankit. (2020). "Is the Time Right for Japan to Become Five Eyes' 'Sixth Eye'?" *The Diplomat*, Aug 14.

Panda, Jagannath. (2021). "AUKUS: Resetting European Thinking on the Indo-Pacific." Edited by Niklas Swanstrom. Institute for Security and Development Policy, 2021. https://isdp.eu/content/uploads/2021/10/AUKUS-Resetting-European-Thinking-on-the-Indo-Pacific-9.11.21.pdf.

Pfluke, Corey. (2019). "A History of the Five Eyes Alliance: Possibility for Reform and Additions." *Comparative Strategy*, Vol. 38, Iss. 4, 302–315, August 19, 2019. DOI: 10.1080/01495933.2019.1633186.

Rogers, D. (2015). "Extraditing Kim Dotcom: A Case for Reforming New Zealand's Intelligence Community?" *Kōtuitui: New Zealand Journal of Social Sciences Online*, Vol. 10, Iss. 1, 46–57, DOI: 10.1080/1177083X.2014.992791.

Thomas, Martin. (2008). "Signals Intelligence and Vichy France, 1940–44: Intelligence in Defeat." *Intelligence and National Security*, Vol. 14, Iss. 1, 176–200, January 2, 2008. DOI: 10.1080/02684529908432529.

Tossini, Vitor J. (2021). "The Five Eyes - the Intelligence Alliance of the Anglosphere." *UK Defence Journal*, May 27, 2021. https://ukdefencejournal.org.uk/the-five-eyes-the-intelligence-alliance-of-the-anglosphere/.

Tréguer, Félix. (2017). "Intelligence Reform and the Snowden Paradox: The Case of France: Article." *Media and Communication*, March 22, 2017. https://www.cogitatiopress.com/mediaandcommunication/article/view/821.

Cross-border Cooperation between Columbia, Peru, and Venezuela: Combatting Illicit Narcotics Trafficking

Laura J. Gender[1]

David J. Kritz[2]

ABSTRACT

The growing instability in the Andes subregion including Colombia, Peru, and Venezuela was exacerbated by the current political landscape, corruption, economic downturns, and conflict escalation, causing increased insecurity along the borders. The lack of a holistic counternarcotics strategy is intensified by the intricate and evolving transnational criminal organization landscape posing a strategic threat to democratic nations in the region. This research analyzed the effects of geopolitics, political conflicts, and resulting transnational criminal activity on cross-border cooperation and intraregional communication between these three nations. A qualitative methodology using descriptive and analytical research methods addressed the central research question: To what extent do the political landscapes in Colombia, Peru and Venezuela effect the exploitation of their porous borders for the flow of illicit narcotics? This article clarifies how national political motivations affect border stability and directly affect the resulting transnational criminal activity and increased illicit narcotics trafficking along the borders.

Keywords: Trafficking; Illicit Drugs; Border Cooperation; Intelligence

Cooperación transfronteriza entre Colombia, Perú y Venezuela: Combate al narcotráfico

RESUMEN

La creciente inestabilidad en la subregión de los Andes, que incluye a Colombia, Perú y Venezuela, se vio exacerbada por el panorama político actual, la corrupción, las crisis económicas y la escalada de conflictos, lo que provocó una mayor inseguridad en las fronteras.

1   Corresponding Author, LGender101@gmail.com

2   david.kritz@mycampus.apus.edu

doi: 10.18278/si.10.1.7

La falta de una estrategia integral antinarcóticos se ve intensificada por el complejo y cambiante panorama de las organizaciones criminales transnacionales, que representa una amenaza estratégica para las naciones democráticas de la región. Esta investigación analizó los efectos de la geopolítica, los conflictos políticos y la consiguiente actividad criminal transnacional en la cooperación transfronteriza y la comunicación intrarregional entre estas tres naciones. Una metodología cualitativa, con métodos de investigación descriptivos y analíticos, abordó la pregunta central de investigación: ¿En qué medida los panoramas políticos de Colombia, Perú y Venezuela afectan la explotación de sus fronteras porosas para el flujo de narcóticos ilícitos? Este artículo aclara cómo las motivaciones políticas nacionales afectan la estabilidad fronteriza e inciden directamente en la consiguiente actividad criminal transnacional y el aumento del tráfico ilícito de narcóticos a lo largo de las fronteras.

Palabras clave: Tráfico; Drogas ilícitas; Cooperación fronteriza; Inteligencia

哥伦比亚、秘鲁和委内瑞拉的跨境合作：打击非法毒品贩运

摘要

安第斯地区（包括哥伦比亚、秘鲁和委内瑞拉）日益加剧的不稳定性因当前的政治格局、腐败、经济衰退和冲突升级而加剧，导致边境地区的不安全感增加。错综复杂且不断发展的跨国犯罪组织格局加剧了整体禁毒战略的缺乏，对该地区民主国家构成了战略威胁。本研究分析了地缘政治、政治冲突及其引发的跨国犯罪活动对这三个国家跨境合作和区域内交流的影响。本研究采用描述性和分析性研究方法（定性方法），探讨了核心研究问题：哥伦比亚、秘鲁和委内瑞拉的政治格局在多大程度上影响了"利用其漏洞百出的边境进行非法毒品流通"一事？本文阐明了国家政治动机如何影响边境稳定，并直接影响由此产生的跨国犯罪活动以及边境地区非法毒品贩运的增加。

关键词：贩运，非法毒品，边境合作，情报

Introduction

Venezuela coupled with the Andean countries of Columbia and Peru are the source of more than 95 percent of the heroin and cocaine available in the U.S., highlighting the strategic importance of this region (Hernandez-Roy and Bledsoe, 2023). Columbia and Peru are two of world's top cocaine producing countries, and Venezuela is a major narcotics transit country, smuggling thousands of pounds of cocaine a year into the U.S. (Seelke, 2023; Johnson, 2012). This Andean sub-region has suffered political instability and social turmoil for the last decade and continues to pose a strategic threat to democratic nations in the region. The growing insecurities in this region were aggravated by political turmoil, corruption at all levels and in all organizations, aggravation of regional relations, and conflict escalation causing increased instability along the borders (Farion et al., 2023). The rollover of all three presidencies and unintended consequences of repressive drug policies on criminal territories/transnational criminal organization (TCO) economic control appear to be contributory causes to not only angst between nations, but poor economies, illicit business expansion, and increased illegal activity (ICG, 2020; Williams, 2024; 1 Americas, 2021; ICG, 2023; Criminality in Peru, 2023; Criminality in Colombia, 2023). The balance of power, regional politics, national defense, illicit trafficking, organized crime, and insecurity remain active, ongoing issues (Rhon and Velez, 2022; ICG, 2020; Jutten, 2024).

The remote areas within the Amazon Forest allow for ideal cultivation of coca plants, the vast expanse of concealed forests and mountainous landscape provide invisibility for illicit activities, and the Amazon River provides smuggling routes and a transit corridor between countries. Weak and ineffective state institutions and the nonexistence of border control and/or border closures have increased insecurity, incentivizing criminal activity. Together, these form ideal conditions for illicit trafficking, resulting in an increase in armed group presence and transnational crime in this region (Ford, 2009; Cardenas and Robbins, 2023; ICG, 2020).

The most recent U.S. National Drug Control Strategy specifies the reduction of illicit narcotic supplies as one of its top priorities (The White House, 2022). Collaboration, synchronization, and coordination between vulnerable drug producing nations and U.S. agencies and task forces through investigative and interdiction efforts is ongoing, albeit challenging. The focus is on disruption of the production, manufacturing, and distribution of illegal drugs with the goal of making these countries independent and self-sufficient partners in the fight against illicit narcotics (The White House, 2022). The U.S. has mutually beneficial partnerships with all three nations over the last few decades, spending billions of dollars in security, training, intelligence, and interdiction support primarily centered on counternarcotics (CN). U.S. organizations involved in these CN strategies include the Department of Defense and

State, DEA, CIA, and the U.S. Agency for International Development. To date, there is no single U.S. CN strategic plan for the Andean region (Johnson, 2012; Jutten, 2024). The lack of a holistic CN strategy is exacerbated by nonexistent intraregional cooperation and intensified by the intricate and evolving TCO landscape. Intraregional cooperation to respond to both trafficking and the traffickers is essentially defunct (Global Trends 2040, 2021).

In 2018, over half of the Union of South American Nations members departed the organization (including Columbia and Peru) resulting in ad hoc bilateral cooperation through individual agreements between countries largely highlighting the lack of overarching regional cooperation with sparse appetite to create any type of regional institution moving forward (Rhon and Velez, 2022). The resulting lack of regional leadership caused a power vacuum that may have triggered the downward spiral of increased foreign armed group presence and TCO criminal activity, especially on the borders (ICG, 2020; Jutten, 2024). Territorial infighting, violence, and internal conflicts created a causal chain or domino effect in the region highlighting corruption, poverty, forced displacement of civilians, and illicit trafficking (Williams, 2024; Jutten, 2024). There are very few international organizations or cross-border agreements in place to enable these countries to connect and collaborate to battle the increasing levels of illicit trafficking traversing these permeable borders at an alarming rate (United Nations, 2023; Meyer, 2023).

Research Question and Purpose

The central research question for this article is: *To what extent do the political landscapes in Columbia, Peru, and Venezuela effect the exploitation of their porous borders for the flow of illicit narcotics?* A key question that was instrumental to address the central question was How does the resulting influence, violence, and criminal activity of TCOs along these borders hamper cross-border and security cooperation? The purpose of this article was to analyze the effects of geopolitics, political conflicts, and resulting transnational criminal activity on cross-border cooperation and communication between Columbia, Peru, and Venezuela. The objectives of this article were to: 1) clarify how national political motivations affect border stability; 2) establish how the resulting transnational criminal activity and increased illicit drug trafficking affects individual nation and regional security; 3) uncover additional central standalone or common factors between these nations that pose barriers to cross-border cooperation to illuminate current border tensions; and 4) drawing upon these underlying problems, present potential recommendations or solutions to improve collaboration and collectively counteract narcotics trafficking between Columbia, Peru, and Venezuela.

Theoretical Framework and Research Design

A review of the issues was accomplished using Causation Theory which permitted the ability to compare causal correlations between actions and events as the theory afforded the opportunity for patterns of causality to come to light through different lenses and diverse causal relationships. The research design included a qualitative research methodology with a multi- method approach including both descriptive and analytical research methods utilizing a systematic search of secondary research data. This provided a holistic account of data using multiple peer reviewed sources and allowed for illumination and interpretation of trends or patterns within the data. Documentation and analysis of all data was accomplished through thematic analysis where the data was coded to determine associations. Content analysis then allowed for theme identification (four primary themes and ten sub-themes) to assist in relationship links.

Primary Themes Within the Literature

Columbia, Peru, and Venezuela continue to pose a strategic threat to democratic nations in the region. The majority of the literature we found indicated that criminal violence and economic downturn are directly related to weak governance and internal political strife caused by a revolving door of politicians and raging public distrust and disapproval revealing evidence of a causal relationship between political turmoil and ongoing border issues. Elevated levels of dysfunction across the political landscape illuminate both contributory and sufficient causes related to the current criminal climate in the border regions

(Beittel, 2023; ICG, 2023; Fabris, 2023; Colombia: Freedom in the World 2024 Country Report, 2024; Stuenkel, 2022). The resulting lack of regional leadership caused a power vacuum that may have triggered the downward spiral of increased foreign armed group presence and TCO criminal activity, especially on the borders (ICG, 2020; Jutten, 2024).

Territorial infighting, violence, and internal conflicts created a causal chain or domino effect in the region highlighting corruption, poverty, forced displacement of civilians, and illicit trafficking (Williams, 2024; Hulswit, 2004; Jutten, 2024). The evidence steers toward several causal relationships (single cause-multiple effect and multiple cause-multiple effect) between the weakened political state and police/military forces and resulting increases in criminal activity in the region and civilian insecurities. The lack of a holistic CN cooperation strategy is glaring and is intensified by the intricate and evolving TCO landscape and intraregional cooperation to respond to both trafficking and the traffickers is essentially defunct (Global Trends 2040, 2021).

Causal Relationships Identified in Literature

The literature presented three qualitative reports/cases that used causal relationships to explain actions and events related to CN efforts. Kleiman (2004) in a Congressional Research Service report indicated connective tissues between illicit narcotics and terrorist threats and indicated five distinct ways in which consumer, transit, and source countries for illicit narcotics could contribute to terrorism: cash supply, creating chaos anywhere between production and consumption, corruption generation amongst officials (military, police, politicians), support for a common infrastructure (pathways and capabilities), and competition for intelligence/enforcement resources (drug or terror targets). Kleinman (2004) further stated the links were "sufficiently clear" enough that CN programs should factor their impacts into decision making (20).

Felbab-Brown (2008) relayed how the poor quality and/or the lack of adequate data on illicit narcotics activities hampers the "robustness of causal inferences" with regard to the effectiveness of CN policies, more specifically due to the lack of baselines (2). Felbab-Brown (2008) then indicates, "Data-informed conclusions could still be drawn" (2). Finally, Piazza (2011) reiterated Kleiman's (2004) five causal links between terrorism and illicit narcotics markets adding that narcotics production and pricing are related/predictors to domestic and transnational terror attacks (297). Piazza (2011) adds that the illicit narcotics trade has compromised security and governments and "… helped to produce an environment that is highly conducive to terrorist movements" (298). Piazza makes a connection between the illegal narcotics trade and funding attacks, adding that terrorists "are well-suited to insert themselves into and benefit from the drug trade" (2011, 298). Finally, Piazza (2011) then insinuated a casual chain relationship

that "Terrorist groups thrive in countries politically destabilized by the drug trade" (298). All three authors agree that weak government, narcotics trafficking, and increased TCO activity are connected.

Illicit Trafficking Narcotics Produced and Distributed

With an increase of over 400 percent from 2014–2020, cocaine production is at record levels with nearly all of the cocaine supplied from Columbia, Peru, and Bolivia (Jutten, 2024). Adaptation of criminal organizations to corruption within law enforcement (LE) entities enabled stricter control of criminal networks that are more diversified. Cocaine supply networks have decentralized and fractured, outsourcing certain duties to specific groups which makes supply chain vendors much more difficult to track by LE (Hernandez-Roy and Bledsoe, 2023; Jutten, 2024). The specific production/distribution for each country illuminates the cooperation of criminal groups for cultivation, production, and distribution of a host of illicit narcotics and goods across porous borders where enforcement is either complicit or nonexistent, essentially laying claim to a causal relationship between corruption and illicit activities.

Other Illicit Trade

Narcotics are not the only illicit items trafficked through established routes across Columbia, Peru, and Venezuela borders. Human and weapons smuggling are commonplace. Weapons from

Europe and Russia are smuggled into Columbia and security forces have been known to sell both their weapons and ammunition on the black market to criminals (Criminality in Venezuela, 2023). Corruption in the region has permitted illicit markets to thrive. Illicit trade in fuel, tobacco products, alcohol, minerals, food, electronics, clothing, and counterfeit goods are prevalent especially along the Columbia and Venezuela borders due to the current economic crisis. Criminal rings engage in fuel theft, illegal mining and logging, and extortion (Criminality in Venezuela, 2023; ICG, 2023; ICG, 2020). There was also an uptick in smuggling significantly cheaper counterfeit medicines from Russia, Cuba, and China over the Columbia border and into Peru (Criminality in Venezuela, 2023; Criminality in Peru, 2024). Corruption, bribery, poverty, and permeable borders have resulted in increased instability and violence along these borders.

Transnational Criminal Organizations (TCO)

TCO and armed group activities in all three countries are rampant with illicit activities rising. Porous borders are a strategic weakness that enables these organized criminal networks to invisibly move through border areas and across borders as criminals exert political, social, and economic control (Criminality in Colombia, 2023; 1 Americas, 2021; Criminality in Peru, 2023). The Amazon River and its tributaries provide smuggling routes and a transit corridor between countries. After the breakup of FARC in 2016, an endless stream of armed groups made their way to the northern countries resulting in steep competition for connections and territory in the illicit narcotics chain. Alliances are skin deep when it comes to controlling narcotics corridors (Criminality in Peru, 2023; Cardenas and Robbins, 2023). The causal relationship of multiple cause-multiple effects through contributory causes is evident as there are several authors who referred to unintended consequences of repressive drug policies on criminal territories and their economic control leading to a plethora of illicit activities (1 Americas, 2021; ICG, 2023; Criminality in Colombia, 2023; Velasco, 2022).

Theoretical Framework

A review of the issues was accomplished using Causation Theory which allowed for data collection and analysis with no preconceived notions. This also permitted the ability to compare causal relationships between the political landscape/turmoil, corruption, TCO/armed group presence, and illicit activity across all three borders in tandem with reviewing outcomes from previous relevant studies. Causality (cause and effect), also known as causation, purports that one action (cause) contributes to another event (effect) where the cause is partly to blame for the effect and the effect is partly reliant on the cause, thereby creating a causal relationship between the two events (Williams, 2024; Hulswit, 2004).

The Four Types of Causal Relationships

Single cause-multiple effect (one action-many effects), multiple cause-single effect (several actions- one effect), causal chain/domino (one effect causes other effects), and multiple cause-multiple effect (several actions-several effects) (Williams, 2024).

Three types of causes include contributory causes (assist in events occurring, but cannot independently generate effects), necessary causes (must exist for event to occur but may not be liable for event), and sufficient causes (what is required for event to occur) (Volchok, 2015).

Gaps in Research

There is a need to conduct further research on the causes and effects of specific actions and events along the borders of these three nations. Although there is sufficient research on the events, we found a paucity of specific evidence of causal relationships between them—only suggestions of correlations. Combined, the evidence and the overarching view of previous researchers suggest the possibility of causal connections between the weakened political state and police/military forces and resulting increases in criminal activity in the region and civilian insecurities. Discovery and documentation of this evidence can assist nations in the region to collaborate and cooperate toward common goals to improve situations or thwart causal triggers. There is also limited evidence

on the topic of intermestic security and the benefits of regional collaboration to address this topic. Finally, there is limited data available on the terrorism nexus associated with the international narcotics trade and the effects of interdiction/eradication efforts on terrorism.

Research Design and Data Analysis Process

A qualitative methodology with a multi-method approach that consisted of descriptive and analytical research were selected to address the central research question. Our research employed a multimethod approach using both descriptive and analytical research. Glass and Hopkins (1984) define descriptive research as, "Gathering data that describes events and then organizes, tabulates, depicts, and describes the data collection" (2). Knupfer and McLellan (2001) states that, "Descriptive research is used as a tool to organize data into patterns that emerge during analysis. Those patterns aid the mind in comprehending a qualitative study and its implications" (1197). Descriptive research was selected for the ability to conduct pattern analysis. This type of research allowed for grouping events and actions into themes and patterns to determine why something happened. This method greatly assisted in addressing the central research question in that patterns developed across all three nations within the thesis that then enabled inferences as to causes and effects of actions and events. It also provided the what, where, and when of the problem posed in the research question effectively describing the problem and situation across the region. Descriptive research through literature reviews allowed a broad look at the situation and circumstances of cross-border cooperation providing the ability to identify trends, categories, and/or characteristics of factors involved.

We also used analytical research to complement the descriptive research method. The Gulf University for Science & Technology (2024) defines analytical research as, "An in-depth study and evaluation of available information in an attempt to explain complex phenomenon. The researcher has to use facts or information already available and analyze them to make a critical evaluation of the material" (1). Valcárcel (2017) states, "Analytical research is used to find the missing link in a study and is focused on the why and on cause and effect" (11168).

This method significantly addressed the central research question as it provided explanations of border situations and enhanced our ability to draw patterns, insights, and conclusions through analysis. Analytical research afforded a comparison of facts and data to assist in the overall evaluation of causal relationships within the data.

Thematic analysis was used to analyze the data where pertinent information was coded using the qualitative analysis software tool Taguette. Over 400 codes were then reviewed for potential themes using content analysis that determined the presence of four primary themes and ten sub-themes that enabled discovery of causal relationships.

Findings & Analysis

Data Collection Procedures

Descriptive research was conducted through the use of the APUS Library search engine and Google Scholar to conduct a systematic search of secondary research data within six databases (JSTOR, Taylor & Francis, ProQuest, Open Access, ABC-CLIO, and Directory of Open Access Journal {DOAJ}) which yielded 43 pertinent studies that met the inclusion conditions. These were supplemented by an additional 30 relevant studies from professional and/or government sources. To be included, literature needed to contain information on Columbia, Peru, and/or Venezuela and match two or more pertinent keywords. See Table 1 for a list of keywords and frequency of use within studies selected. Descriptive research allowed a broad look at the situation and circumstances of cross-border cooperation providing the ability to identify trends and/or characteristics of factors involved. Analytical research was then used to complement the descriptive research method where each study was critically reviewed for possibly causal contributions, then initial patterns were identified to complete a comprehensive picture derived from data collected. Finally, an evaluation on each study was accomplished to interpret causal links between events and actions. Analytical research afforded a comparison of facts and data to assist in the overall evaluation of causal relationships within the data.

Table 1. Keywords and Frequency

Keyword	Frequency	Keyword	Frequency
Government	92	Eradication	33
Illicit	86	Petro	32
Security	82	Congress	27
Criminal	81	Terrorism	23
Politics/political	79	Police	22
Military	77	Insecurity	19
Border	67	Interdiction	18
Armed Group	57	Counternarcotics	15
Corruption	55	Corrupt	10
Maduro	54	Boluarte	9
Drugs/narcotics	39	Instability	9
FARC	38	Weak	9
Violence	38		

Data Analysis Process

A systematic documentation and analysis of all data was accomplished through thematic analysis. The qualitative analysis software tool Taguette was used to methodically code the data to determine patterns and associations which resulted in the creation of 410 codes. Content analysis then allowed for specific themes and concepts to be determined to assist in relationship correlation. This resulted in the identification of four primary themes and an additional ten subthemes nested within them.

Findings

An extensive review of 70+ studies and professional documents brought forth both observations and truths regarding the effects of political decision-making on porous borders in the Andean region that have incentivized criminal activity. The overarching view of the majority of studies identified causes and effects of actions and events that have exacer-

bated the current situation within the border regions. Analysis of the research identified four principal themes: political turmoil, corruption, insecurity/instability, and TCO activities and ten sub-themes that were illuminated as having directly or indirectly contributed to the primary themes.

Political Turmoil

Weak democratic institutions and governmental flaws are recognized as a consistent problem in the Andean region, incentivizing crime, impunity, and violence in the region. Political instability allows armed groups and TCOs to operate freely. Jutten (2024) states, "Political instability experienced in Andean countries in recent years goes hand in hand with a worsening public security situation and rising cocaine supplies" (1). The results show that both governmental decision-making and the infiltration of TCOs into governmental offices and organizations intensifies the political turmoil in these nations. The rollover of all three presidencies

Table 2. Themes, Sub-Themes, Tertiary Connections, and Other Effects

Theme	Sub-theme	Tertiary Connections	Other Effects
Political Turmoil	TCOs in government		
	Government decision-making	Lack of cross-border cooperation	Strained bilateral relationships
			Minimal bilateral cooperation
Corruption	Military		
	Police		
	Government officials		
Insecurity / Instability	Border instability		
	Civilian insecurities		
	Interdiction Efforts	Eradication efforts	
		Alternative cultivation strategies	
TCO Activity	Armed group activity		
	Instability nexus		
	Illicit trafficking	Violence nexus	
		Terrorism nexus	
		Financial nexus	
		Illicit trade	
		Narcotics trafficking	

was recognized as a contributory cause to not only angst between nations, but flying accusations, stunted economic growth, illicit business expansion, and an uptick in illegal border crossings and activity. The International Crisis Group (2023) indicates, "Weak democratic institutions, high levels of corruption, and extreme inequality have made Latin America fertile ground for organized crime" (3).

Corruption

Corruption is rampant in all three nations infiltrating all levels of government, LE, and the military. In the political realm, criminal organizations have infiltrated governmental decision-making rendering judicial systems useless as impunity for criminals remains elevated. Arsht (2023) notes, "Beyond the escalating violence, recent high-profile cases of corruption have raised concerns around the linkages between political parties and campaigns with illicit funds tied to criminal organizations, further threatening democratic institutions and the rule of law" (8). Police and military forces plagued by low wages, poor training, and increased expectations of patrolling extremely violent cities often participate in or enrich themselves by colluding with TCOs through a variety of illicit activities. The International Crime Group (2023) indicates, "New crime hubs have emerged in areas that offer strategic benefits to drug traffickers and enable novel connections to be forged among transnational outfits, local gangs, and corrupt officials in courts, prisons, and police forces" (3).

Insecurity/Instability

Violence is frequently used for political purposes which undermines safety and stability and eventually the state's authority. Dramatic security deterioration continues to occur, especially in the border regions as TCOs and armed groups increase territorial boundaries and power as these groups exert influence over border populations. 1 Americas (2021) states, "Around 25 percent of the urban population in Latin America is poor, and widespread informality and unemployment (especially for youth) provide the perfect terrain for criminal gangs and illicit activities to thrive" (39). Facts show that civilian insecurities abound through extremely high poverty rates, minimal access to basic services/healthcare, poor infrastructure, and daily threats from TCOs regarding coca cultivation. A surprising cause of insecurity was failed interdiction and forced eradication efforts. Felbab (2008) indicates, "Eradication frequently greatly impoverishes poor farmers strongly dependent on illicit crop cultivation for basic livelihood. The economic hardship experienced by targeted farmers has led to political unrest, violent protests, and destabilization of governments in source countries" (14).

TCO Activities

Coca cultivation rose dramatically as paramilitaries and TCOs increased their presence and expanded territories on the borders resulting in permissive environments that allowed for growth, influence, and increased revenue through political protection and impunity. Arsht (2023) noted, "Concerns over crime, violence, and insecurity have risen in recent years as a result of organized crime linked to drug trafficking. TCOs have evolved and taken advantage of regional instability to leverage their well-established networks to adapt to new market dynamics and diversify their services" (3). Data demonstrated TCO activity included illicit narcotics production and distribution, illicit trade, and a range of criminal activity through violence, threats, and exploitation of people and environments. Results indicated illicit trafficking caused instability and chaos weakening government and LE's ability to intervene.

Findings Conclusion

The findings revealed the pervasiveness of political turmoil and corruption in the Andean region has either a direct or indirect effect on stability and security within Columbia, Peru, and Venezuela. Weak governance has allowed for increased TCO activity along porous borders which increased both border instability and civilian insecurities resulting in an uptick in illicit trafficking and criminal activity. The results show interconnectivity between actions and events in the region suggesting causal relationships that have a detrimental effect on cross-border cooperation.

Analysis

This analysis supports the theory that political decision-making and turmoil has a direct effect on the exploitation of porous borders, increased TCO activity, and the lack of cross-border cooperation between Colombia, Peru, and Venezuela. The majority of the literature indicated the criminal violence and economic downturn are directly related to weak governance and internal political strife caused by a revolving door of politicians and raging public distrust/disapproval revealing evidence of a causal relationship between political turmoil and ongoing border issues. Elevated levels of dysfunction across the political landscape illuminate both contributory and sufficient causes related to the current criminal climate across the border regions emphasized through four primary themes within the research (Beittel, 2023; ICG, 2023; Fabris, 2023; Stuenkel, 2022).

Political Turmoil

Columbia's President Petro is facing administration dissatisfaction and growing insecurity, Peru's President Boluarte currently holds an 84 percent disapproval rating that is contributing to a failing political system, unrest, and polarized extremes, and Venezuela's authoritarian President Maduro (currently in the midst of another contested reelection) faces international isolation and continues to be a regional source of instability. All three new presidents are facing widespread opposition and questions of democratic legitimacy in their individual countries. The result is instability, protests, extensive corruption, collapsing institutions, and spiraling crime resulting in increased armed

group violence and migration issues throughout the region (Criminality in Peru 2023; Criminality in Venezuela 2023; ICG 2020).

The common threads that run through these three governments are lack of transparency in elections and decision-making, weak and divided governance, distrust, dishonesty, and institutional dysfunction. These threads typically result in causal relationships associated with public approval. Disapproval rates are soaring as a domino effect of political bantering and meandering has resulted in extremely high levels of corruption and national and regional instability. The International Crisis Group (2024) indicates, "Public disaffection with leaders, state institutions, political parties, and democracy in general are running at record highs" (36).

The infiltration, penetration, and linkage of TCO members into the political realm and processes is evident with FARC as a prime example, formerly the largest guerilla group in COL. In 2016, when 10,000 FARC members disbanded after the Peace Accord, the group was reincarnated as their own political party (Comunes Party) consisting of one thousand members eventually winning 10 seats in Congress and who, to this day, vow to represent former FARC members (1 Americas, 2021, 61). Corruption between political authorities and armed groups is rampant as government officials routinely accept bribes for votes and grant impunity, seeing high-level politicians directly accused of both collaborating with TCOs and being involved in narcotics trafficking. The government's decision to negotiate with guerillas directly resulted in the formation of 30 additional dissent groups who quickly took FARC's place in controlling territory, drug and human trafficking, and illicit trade on the borders illuminating a causal chain between political decision-making and unintended consequences of increased armed group activity and involvement. Arsht (2023) indicates, "Transnational organized crime, and its penetration of state institutions, represents one of the fastest-growing threats to global governance today" (3).

The evidence suggests that political turmoil is directly associated with disarray, increased corruption, instability, and increased TCO activity. Political instability has been ongoing in the Andean region for over a decade exhibiting a causal relationship between political disruptions and rising insecurity in the region. These results build on existing evidence from literature that studied the political elections and resulting turmoil over the last decade drawing the same conclusions of causal relationships between these events and results. Rhon and Velez (2022) stated, "Regionally we have learned nothing from the lessons of the history and of the poor results of democratic governance regarding cooperation in defense and security" (161).

Corruption

Corruption is a significant concern that poses a domino effect throughout the three countries leading to inept, illegitimate, and weak governance as well as collusion and complicity between government institutions and criminal networks and their illicit activities. Instability in this region caused by weak institutions and high levels of inequality have provided fertile ground for TCOs to take hold inducing a noticeable uptick in illicit narcotics trafficking distinctly connected to high levels of corruption (Ramsey and Smilde, 2020; ICG, 2023; Rincon and Sultan, 2024). Combined, the evidence and the overarching view of authors steer toward a causal relationship between the weakened governments and state/police forces and both increased criminal activity in the region and civilian insecurities. Hassan (2024) states, "Corruption is a major factor in the deterioration of public institutions, deficient public services, and environmental destruction" (4).

As identified throughout the research, corruption was reported across the three countries in government, in LE, and in the military. An explanation of corruption within the government was discussed in a previous section within this article however, it is worth noting that corruption within both the police and the military would not transpire without the complicity of government officials. Ramsey and Smilde (2020) state, "There is no question that organized crime and corruption have flourished in the midst of Venezuela's crisis. There is credible evidence of many officials' corruption, involvement in illicit activity, kickbacks, and patronage schemes" (4). The data suggests these three governments have enabled and encouraged illicit activities as most are complicit and sympathetic to armed group activities with several high-level politicians providing safe haven for criminals and drug traffickers. Complicit officers in both military and police and tolerant local supporters together with permeable borders continue to offer impunity to criminals producing high-level security concerns for these bordering nations. Beittel indicates, "The Venezuelan military and other authorities have been complicit in the ELN's illicit activities in many instances … and has provided the ELN with safe haven from the Colombian military, allowing the ELN to expand into Venezuelan territory and criminal markets" (4).

Cocaine trafficking networks are too large to escape detection and infiltration by LE and/or military forces resulting in extensive intelligence networks that include both LE and TCOs to ensure protection of smuggling routes and laboratories and safeguard against major player incarceration (Vélez, 1995; Jutten, 2024; Ramsey and Smilde, 2020; ICG 2022). Border monitoring/controls and arrests for criminal activities are hampered by complicit involvement and turning a blind eye resulting in permeable border corridors for illicit trafficking. Data suggested armed groups have penetrated the military ranks in order to gain valuable intelligence on operations, frequently targeting former military. Providing high payouts, security forces are quick to join ranks or accept the cash for informing on colleagues or redirecting patrols. The International Crisis Group (2022) stated, "Armed groups also try to forge relationships with retired soldiers, who understand military networks and can more easily penetrate them" (21). Conversely, police/security forces are typically intimidated, undermined, and threatened into assisting drug traffickers. However, these forces can also be willing participants due to low wages, extensive discretionary power, and "easy money" offered by traffickers for safe passage through border regions. These events led to the distrust of government, police, and military legitimacy and effectiveness in the public eye. A causal domino effect of trafficking, police/military brutality and corruption, and the lack of security along borders contributes to the perpetuation of insecurity in the region. This research has verified and concluded the existence of a causal relationship between both political decisions and corruption, and corruption and insecurity/increased illicit TCO activity.

Insecurity/Instability

Border instability, civilian insecurities, and interdiction efforts continue to exacerbate instability in the region as the data suggested all three were causes with direct effects on instability and insecurity in the region. Border instability is caused by a variety of actions, but primarily TCO activity and control of the borderland populations. These populations are subject to illicit narcotic activity fallout including becoming targets of local armed groups if suspected of assisting the police or military, or targets of paramilitary forces that suspect opposition to their activities. A significant amount of research showed these acts are tolerated by civilians as many rely on coca cultivation to support their families. A considerable number of TCO members reside on the borders and most civilians are subject to border wars as drug trafficking groups frequently fight for territory and trafficking corridors. This results in trickle-down effects to borderland civilians as their coca leaf purchasers routinely change. Crimes are not constrained by individual borders but instead are aided by gaps in police or military enforcement that have increased insecurity, incentivizing this criminal activity. Traffickers take advantage of natural infrastructure provided within the rural border regions as well as the absence of state enforcement intensifying the insecurities. The fact that border regions are mostly rural, many civilians feel a sense of state abandonment as security forces rarely patrol borderland regions effectively leaving the armed groups to maintain security and control. These revelations heighten the requirement for a security and counternarcotics strategy that promotes increased LE and military involvement within the border regions.

A surprising and unknown fact emerged within the data regarding intermestic security. This term denotes the relevance and interrelation between international and domestic security concerns. Regional security issues and multinational threats can pose challenges and internal security concerns creating a conflict between national and international politics. Intermestic security threats can include internal vi-olence/conflict, paramilitaries, political instability, narcotics trafficking, organized crime, and instability from other countries (Bragatti and Weiffen, 2023). Rhon and Velez (2022) state, "Nations experiencing intermestic security concerns may use this as an opportunity to join forces to maintain bilateral relations to resolve border criminality" (156). Research highlighted both the dangers of this regional and global concern and insinuated it could bring nations together to join forces against the threats. This revelation in literature is important because of its innate ability to bring together nations who are experiencing the same threats into improved or increased bilateral, regional, or cross-border cooperation.

Civilian insecurity reigns throughout each country but is worse in the borderland regions. Instability caused by weak institutions and high levels of inequality have provided fertile ground for TCOs to take hold inducing a noticeable uptick in illicit narcotics trafficking. Rubiano et al. (2018) stated, "Today, almost 42 percent of Colombia's rural population falls beneath the poverty line. Coca cultivation in Colombia is fueled by the vast levels of poverty that exist in the country" (5). Lack of enforcement has caused innumerable insecurity issues for populations in the borderlands where very small percentages of the population have access to safe water sources and sanitation with a significant decline in public services due to corruption and undue pressures/influences in the region. These primarily stem from cultural, gender, and socioeconomic disparities. Criminality in

Columbia (2023) indicated, "The geography of the country has also made establishing the effective presence of law enforcement bodies difficult in remote areas, where criminal groups have managed to exert social, economic and political control" (3).

The research suggests that the causal relationship between civilian insecurities and both the corruption it causes and resulting TCO activity/control along the borders directly speaks to exploitation of porous borders. Interdiction efforts across the three nations have been sporadic at best in recent years. Plan Colombia introduced forced eradication as a repressive drug policy as part of its interdiction efforts. These policies included increased military presence and extermination of illicit coca crops. An unexpected outcome from research was knowledge of the complete and utter failure of eradication efforts within Columbia. Not only did these efforts alienate the borderland populations, but they had inadvertent consequences for state forces (police and military) as criminal groups frequently retaliated against these efforts. Eradication efforts impoverished farmers as these efforts destroyed both illicit and licit crops as well as completely contaminating water sources used for drinking and field irrigation. An unfortunate side effect was the resulting push of farmers toward illicit crop cultivation due to their increased economic dependence on these crops and intensified bonds formed with armed groups because of these failed efforts. These interdiction efforts failed to stop or even slow criminal groups' ability to shift

areas and/or replant crops after eradication highlighting the ability of drug traffickers to adapt methods and routes around eradication efforts. Alternative livelihoods (incentivizing farmers to grow licit crops) also failed as programs were poorly funded, ill-conceived, and not focused on long term. Consequently, the inability for drug-producing countries to create and fund alternative eradication efforts (bilaterally, regionally, or nationally) has allowed TCOs and armed groups to increase their control not only over the region but over borderland populations offering alternative means of support.

TCO Activity

Armed group activity increased for decades in the Andean region. Rhon and Velez (2022) indicated, "For over 40 years that subregion has displayed an unwelcome and complex interdependence, a product of the massive growth of transnational organized crime that represents a permanent strategic threat for the states concerned" (153). The data suggests permissive environments and porous borders have allowed criminal actors and illegal activities to thrive increasing their influence over borderland communities and their grip over political organizations that afford them protection and impunity. The research also indicated chaos caused by TCOs and illicit narcotics markets directly contribute to political instability and undermine domestic security. The literature presents a plethora of criminal activities that supports narcotics trafficking including violent crimes, extortion, torture, child recruitment, kidnapping, sexual violence/human rights violations, human trafficking, and murder (Country Reports on Human Rights Practices: Colombia, 2022; Colombia: Freedom in the World 2024 Country Report, 2024; Beittel, 2023). Two sub-themes emerged from the research indicating instability nexus's (violence, financial, terrorism) and illicit trafficking have direct correlations with TCO activity.

There were distinct connections made through the studies that showed causal relationships between TCOs and violence, TCOs and finances, and TCOs and terrorism. TCO activity along the borders is highlighted by extreme violence that is typically used to control borderland populations in an effort to gain and control territory and crops and infighting is typical as groups wrangle for control over trafficking corridors. Violence is aimed at civilians to both keep them in line and to punish them for assumed indiscretions including informing and reporting on them to authorities. TCO members also frequently attack police and military offices. 1 Americas (2021) stated, "Coca crops and cocaine production remain important drivers of violence, as the most important revenue source for illicit criminal and terrorist groups, as well as an income source for marginalized rural communities" (58). The evidence suggests violence is intricately intermingled with the financial/revenue side of illicit narcotics trafficking. Most violence is instigated over competition for territory and illegal economies. Studies indicated that armed groups traffic illic-

it goods strictly for the profits and control of the market and consequently, this article demonstrates the same essential point. Finally, several studies indicated a correlation between TCO activity and terrorism due to an influx in terrorist group involvement in illicit narcotics trafficking, noting these groups thrive in politically destabilized countries. Furthermore, the evidence showed drug trafficking markets are easily exploited by terrorist organizations with minimal effort and investment. Causal connections between these nexus's and TCO activity are evident in the association of violence, money, and terrorism with illicit trafficking on the borders.

Illicit trafficking is inclusive of illicit narcotics and a plethora of other items that are illicitly traded to include human and weapons trafficking, counterfeit goods, tobacco, food, clothing, electronics, fuel, and alcohol among others. Cocaine production increased over 400 percent in a six-year period and remains at record levels (Jutten, 2024). Research indicates adaptation of criminal groups to police and military corruption potential that enabled stricter control of diversified networks that have decentralized and outsourced. Likewise, the specific production/distribution for each country illuminates the cooperation of criminal groups for cultivation, production, and distribution of a host of illicit narcotics and goods across porous borders where enforcement is either complicit or non-existent, subsequently laying claim to a causal relationship between corruption and illicit activities. 1 Americas (2021) indicated, "Competition over former

FARC-controlled areas, coca cultivation, cocaine production and distribution corridors, illegal mining, and extortion continued to underpin ongoing dynamics of violence" (55). The evidence shows a clear causal relationship between illicit narcotics trafficking, TCO activity, violence, corruption, and illicit trade.

Analysis Conclusion

Political landscapes and decision-making significantly affect TCO's exploitation of porous borders within Columbia, Peru, and Venezuela. The data suggested weak governance, dysfunction, divided parties, and distrust have an inextricable connection directly contributing to deteriorating instability and insecurity in the region. Both corruption and instability cause disarray and are significant factors in the deterioration of security on the borders. The evidence suggests TCO groups have complete autonomy in the use of trafficking corridors and permeable borders for increased illicit trafficking due to complicity and collusion of the government, LE, and military effectively hampering cross-border security and cooperation.

Recommendations

Regional and cross-border cooperation is paramount to maintaining security and stability within this Andean subregion. Distrust and lack of commitment toward common goals prevents these nations from communicating and collaborating to resolve shared threats. Below are recommendations these nations could make to advance shared cooperation:

1) Open lines of communication between nations for training, consultations, and routine information sharing including all levels of government (especially local levels) to build trust

2) Determine potentialities and commonalities toward benefits to resolving shared threats

3) Divide cooperation into three levels: short-term (mutual agreements or pre-policy), mid-term (commitment to perform and execute), and long-term (finalize policy and institutionalization of agreements)

4) Commit to improved intelligence and information sharing and capacity building by seeking sustainable solutions to threats that are of mutual interest

Future Research

Future research could continue to investigate causal relationships of actions and events on border regions in the Andean region. Further examination of counternarcotics strategies and plans inclusive of bilateral or regional cooperation would benefit leaders in the region enabling them to have a broader view of how cross-border security and cooperation can increase through common goals and efforts. Deeper knowledge of intermestic security could increase our knowledge of how it affects countries in the region and how these threats can be better conquered through regional vice national efforts. Finally, further research into the terrorism nexus associated with the international narcotics trade and the effects of interdiction/eradication efforts on terrorism is needed to better understand the regional terrorism threat and how increased TCO presence and illicit narcotics trafficking affects terrorism as a whole.

The goal for these nations should be to improve trust, communication, and cooperation toward shared concerns and threats in the region emphasizing cross-border cooperation and agreements. Inter and intraregional as well as international organizations can and should be engaged to improve communication, border control, policing, capacity building, and shared intelligence/information/resources to improve security and stability along the borders. Disengaging, disincentivizing, and discouraging TCO presence and activities is paramount to maintaining peace and security in the region.

Conclusion

The U.S. considers reduction of the out-of-control drug supplies arriving from the Andean region as a top priority and efforts to counter these threats start with regional cooperation. Columbia, Peru, and Venezuela had historically weak and ineffective state institutions, encouraging armed groups to come to this region for the sole purpose of taking control of Amazon narcotics pathways. Crimes are unconstrained by individual borders and are aided by gaps in police or military enforcement with increased insecurity that incentivizes criminal activity.

The rollover of all three presidencies and unintended consequences of repressive drug policies on criminal territories/TCO economic control are said to be contributory causes to a variety in events along the border region. The majority of the literature agreed that the criminal violence and economic downturn are directly related to weak governance and internal political strife caused by a revolving door of politicians and raging public distrust/disapproval revealing evidence of a causal relationship between political turmoil and ongoing border issues. Elevated levels of dysfunction across the political landscape illuminate both contributory and sufficient causes related to the current criminal climate in the border regions.

The causal chain of continued police/military brutality, armed group infighting, and lack of security along the border sparking increased illicit trafficking again highlights both contributory and sufficient causes for the influx of TCO activity along the border regions. Corruption is a significant concern and in and of itself poses a domino effect throughout the country leading to inept, illegitimate, and weak governance as well as collusion and complicity between government institutions and criminal networks and their illicit activities. Border disputes are a primary reason for hampered CBC between nations and the research showed minimal effort and reporting on CBC with regards to illicit narcotics trafficking. Thwarted efforts to plan, engage, and enact CN plans shows sufficient causes for continued advancement of these TCO criminal activities and their spillover domino effects in border regions that continue to display causal relationships with trickle-down effects to civilian populations.

This research centered on analyzing to what extent political landscapes have an effect on porous borders for the flow of illicit narcotics and how the resulting TCO activities affect cross- border and security cooperation? We analyzed the effects of geopolitics, political conflicts, and resulting transnational criminal activity on cross-border cooperation and communication between Columbia, Peru, and Venezuela. Identification of four primary themes (political turmoil, corruption, insecurity/ instability, and TCO activity) revealed suggested intertwined causal relationships that continue to intensify the current fragile border situations. The research both discovered and explained causal relationships between events that effect collaboration on all sides of the borders which allowed for the illumination of plausible factors and causes for the lack of regional cooperation.

The findings revealed the pervasiveness of political turmoil and corruption in the Andean region has either a direct or indirect effect on stability and security within Columbia, Peru, and Venezuela. Weak governance has allowed for increased TCO activity along porous borders which increased both border instability and civilian insecurities resulting in an uptick in illicit trafficking and criminal activity. These results directly address the research question of intertwined political turmoil allowing for influxes in illicit narcotics trafficking across permeable borders that subsequently both increase insecurity and cause extreme lapses in regional cooperation. This research contributes to and builds upon existing research by addressing the research gap of the lack of cause and effects evidence of specific actions along the regional borders. It also provides evidence of causal relationships between them by clarifying cause and effect triggers in the region.

Laura Gender holds an MS in Intelligence Studies. Her primary area of research includes counter narcotics operations and cross-border cooperation and how political landscapes affect porous borders and the subsequent flow of illicit narcotics and goods. Highlights from her research include the effects corruption, regional insecurities/instabilities, and transnational criminal activity have on civilian populations and the countries and regions as a whole.

David J. Kritz holds a Doctorate in Business Administration and an MS in International Relations. He is a Full Professor and the Assistant Department Chair of Intelligence Studies in the school of Security and Global Studies for the American Military University, adjunct professor for the Center for Intel & Security Studies

Program at Ole Miss, an adjunct professor for SUNY Empire State College, the assistant editor for The American Intelligence Journal, and an official reviewer for the Journal of Leadership Education. He is a retired from the U.S. Air Force as an intelligence officer and the key position of MSSI Program Director at the National Intelligence University.

References

"1 Americas." (2021). Armed Conflict Survey 7 (1): 35–85. https://doi.org/10.108 0/23740973.2021.1974247.

Arsht, Adrienne. (2023). "Advancing US-Colombia Cooperation on Drug Policy and Law Enforcement." Atlantic Council's US-Colombia Advisory Group. Atlantic Council. November 29, 2023. https://issuu.com/atlanticcouncil/docs/report_us-colombia_counternarcotic_4.

Beittel, June S. (2023). U.S.-Colombia Security Relations - Future Prospects in Brief. Congressional Research Service (CRS) Report No. R47426. Washington, D.C.: Congressional Research Service. https://crsreports.congress.gov/product/pdf/R/R47426.

Bragatti, Milton Carlos, and Brigitte Weiffen. (2023). "The Deterioration of South America's Security Architecture: From Cooperation to Coexistence?" International Relations (London). https://doi.org/10.1177/00471178231195246.

Cardenas, Juan Diego and Seth Robbins. (2023). "Expanding Drug Trafficking on Peru's Borders with Colombia and Brazil." InSight Crime. Igarape Institute. August 8. https://insightcrime.org/investigations/expanding-drug-trafficking-peru-colombia-brazil- border/.

"Colombia: Freedom in the World 2024 Country Report." (2024). Freedom House. https://freedomhouse.org/country/colombia/freedom-world/2024.

"Country Reports on Human Rights Practices: Colombia." (2022). U.S. Department of State. U.S. Department of State. https://www.state.gov/reports/2022-country-reports-on-human- rights-practices/colombia/.

"Criminality in Columbia." (2023). Global Organized Crime Index. https://ocindex.net/2021/country/colombia.

"Criminality in Peru." (2023). Global Organized Crime Index. https://ocindex.net/2023/country/peru.

"Criminality in Venezuela." (2023). Global Organized Crime Index. https://ocin dex.nct/2023/country/vcnezuela.

Fabris, Martino. (2023). "Colombia's Revolutionary Anti-Drug Plan: Break-throughs and Challenges Ahead." The New Global Order. December 22. https://thenewglobalorder.com/world-news/colombias-revolutionary-anti-drug-plan-breakthroughs-an-challenges-ahead/.

Farion, Oleh, Dmytro Kupriyenk, Yurii Demianiuk, and Felbab-Brown, Vanda. (2008). "Counternarcotics policy overview: Global trends & strategies." Washington, D.C.: Brookings Institute.

Ford, J. T. (2009). Drug Control: U.S. Counternarcotics Cooperation with Venezuela Has Declined.

Glass, G. V. and Hopkins, K. D. (1984). Statistical Methods in Education and Psychology, 2nd Edition. Englewood Cliffs, NJ: Prentice-Hall.

"Global Trends 2040." (2021). Office of the Director of National Intelligence. March 2021. https://www.dni.gov/index.php/gt2040-home/gt2040-5-year-regional-outlooks/latin-america.

Hassan, Tirana. (2024). "Peru: Events of 2023." Human Rights Watch. https://www.hrw.org/world-report/2024/country-chapters/peru.

Hernandez-Roy, Christopher, and Rubi Bledsoe. (2023). "Building Barriers and Bridges: The Need for International Cooperation to Counter the Caribbean-Europe Drug Trade." Center for Strategic and International Studies (CSIS). July 14, 2023. https://www.csis.org/analysis/building-barriers-and-bridges-need-international-cooperation-counter-caribbean-europe-drug

Hulswit, Menno. (2004). Causality and Causation: The Inadequacy of the Received View. "A Short History of 'Causation.'" University of Nijmegen. *SEED Journal*, Vol 4, no. 2: 3–21.

International Crisis Group (ICG). (2020). "State, Crime and Poverty along the Border." Disorder on the Border: Keeping the Peace between Colombia and Venezuela. http://www.jstor.org/stable/resrep31452.5.

International Crisis Group (ICG). (2022). "Trapped in Conflict: Reforming Military Strategy to Save Lives in Colombia." International Crisis Group (ICG). September 27, 2022. https://www.crisisgroup.org/latin-america-caribbean/andes/colombia/95-trapped-conflict- reforming-military-strategy-save-lives.

International Crisis Group (ICG). (2023). "Latin America Wrestles with a New Crime Wave." International Crisis Group. https://www.crisisgroup.org/latin-america-caribbean/latin-america-wrestles-new-crime-wave.

Johnson, Jr., Charles Michael. (2012). Counternarcotics Assistance: U.S. Agencies Have Allotted Billions in Andean Countries, but DOD Should Improve Its Reporting of Results. GAO- 12-284. Washington, D.C.: Government Accountability Office.

Jutten, Marc. (2024). "EU Cooperation with Latin America: Combating Drug Trafficking in the Andean Region." Policy Commons. European Parliamentary Research Service. April 22. https://policycommons.net/artifacts/12266675/eu-cooperation-with-latin-america/13163003/.

Kleiman, Mark A. R. (2004). Illicit Drugs and the Terrorist Threat: Causal Links and Implications for Domestic Drug Control Policy. Congressional Research Service Report for Congress, RL32334. Washington, D.C.: The Library of Congress.

Knupfer, Nancy Nelson, and Hilary McLellan. (2001). "Descriptive Research Methodologies." The Association for Educational Communications and Technology (AECT). August 3, 2001. https://members.aect.org/edtech/ed1/41/index.html.

Meyer, Peter J. (2023). U.S. Foreign Assistance to Latin America and the Caribbean: FY2024 Appropriations. Congressional Research Service Report No. R47721. Washington, D.C.: Congressional Research Service. https://crsreports.congress.gov/product/pdf/R/R47721.

Piazza, James A. (2011). "The Illicit Drug Trade, Counternarcotics Strategies and Terrorism." Public Choice 149 (3/4): 297–314. https://doi.org/10.1007/s11127-011-9846-3.

Ramsey, Geoff, and David Smilde. (2020). "Beyond the Narcostate Narrative: What U.S. Drug Trade Monitoring Data Says About Venezuela." WOLA Advocacy for Human Rights in the Americas. https://www.wola.org/analysis/beyond-the-narcostate-narrative-venezuela- report/.

Rhon, Renato Rivera, and Fredy Rivera Velez. (2022). "Chapter 9/South America Under the Pendulum. Bilateralism, Intermestic Security, and the Return of Old Practices." Essay. In Regional and International Cooperation in South America After COVID. DOI: 10.4324/9781003230403-9:148–65.

Rincon, Ricardo, and Ann Sultan. (2024). "Top Five Takeaways on Corruption Perception in Latin America: Insights from Professionals Across the Region." International Bar Association. June 9, 2024. https://www.ibanet.org/top-five-take aways-on-corruption-perception-in- latam.

Rubiano AM, Muñoz JHM, Estebanez G, Sanchez AI, Puyana JCJ, Puyana JC. (2018). Drugs, Violence and Trauma in the Colombian Context: A Health Care Point of View of a Human Rights Challenge. Panam J Trauma Crit Care Emerg Surg 2018; 7(2):158–163.

Seelke, Clare Ribando, (2023). Venezuela: Political Crisis and U.S. Policy. Congressional Research Service Report No. IF10230. Washington, D.C.: Congressional Research Service. https://sgp.fas.org/crs/row/IF10230.pdf.

Stuenkel, Oliver. (2022). "The Greatest Risk Facing Colombia and Its New Leftist President." Carnegie Endowment for International Peace. August 11. https://carnegieendowment.org/posts/2022/08/the-greatest-risk-facing-colombia-and-its-new-leftist-president?lang=en.

The White House. (2022). "Fact Sheet: White House Releases 2022 National Drug Control Strategy That Outlines Comprehensive Path Forward to Address Addiction and the Overdose Epidemic." Last modified April 21, 2022. https://www.whitehouse.gov/briefing-room/statements-releases/2022/04/21/fact-sheet-white-house-releases-2022-national-drug-control-strategy-that-outlines-comprehensive- path-forward-to-address-addiction-and-the-overdose-epidemic/.

United Nations: Meetings Coverage and Press Releases, Security Council Adopts Presidential Statement on Transnational Organized Crime, Urges States to Ramp Up Cooperation, Laws, Borders to Combat It, 2023, https://press.un.org/en/2023/sc15516.doc.htm.

Valcárcel, Miguel. (2017). "Usefulness of Analytical Research: Rethinking Analytical R&D&T Strategies." Analytical Chemistry 89 (21): 11167–72. https://doi.org/10.1021/acs.analchem.7b03935.

Velasco, Marcela. (2022). "Drug Trafficking in Colombia Undermines the Foundations of Indigenous Autonomy." International Work Group for Indigenous Affairs (IWGIA). May 2. https://www.iwgia.org/en/news/4756-drug-trafficking-in-colombia-undermines-the- foundations-of-indigenous-autonomy.html.

Vélez, Hernando Wills. (1995). "Effects of the War on Drugs on Official Corruption in Colombia." Effects of the War on Drugs on Official Corruption in Colombia. Thesis, Monterey, Calif: Naval Postgraduate School. Naval Postgraduate School.

Volchok, Edward. (2015). "Causal Marketing Research." Three Levels of Causation. The City University of New York (CUNY). August 22. http://media.acc.qcc.cuny.edu/faculty/volchok/causalMR/CausalMR3.html.

Williams, Owen. (2024). "Cause and Effect." Department of English. University of Pennsylvania. https://www.english.upenn.edu/graduate/resources/teachweb/owcause.html.

Natural Language Processing for the Intelligence Analysis Process

Brandon Morad[1]

ABSTRACT

This study aims to explore the potential benefits and challenges of integrating Natural Language Processing (NLP) into the U.S. Intelligence Community's (IC) intelligence analysis process. Within the current threat environment taking advantage of this specific Artificial Intelligence (AI) technology is crucial for maintaining international superiority. A qualitative content analysis methodology will be used to answer the research question: should we introduce NLP into the intelligence analysis process? This type of study will enable the comparison of NLP use cases across various fields that conduct data analysis to objectively identify how NLP can improve the practice of intelligence analysis. The chosen lens and theoretical framework for this study provides its own definition of the intelligence analysis process and views NLP as a mechanism to augment that discipline. The aim is to fill a noticeable gap in existing literature and offer valuable insights and implications of adopting NLP within the intelligence analysis process while considering the limitations and biases associated with qualitative studies.

Keywords: Natural Language Processing; U.S. Intelligence Community; Artificial Intelligence; intelligence analysis

Procesamiento del lenguaje natural para el proceso de análisis de inteligencia

RESUMEN

Este estudio tiene como objetivo explorar los beneficios y desafíos potenciales de integrar el Procesamiento del Lenguaje Natural (PLN) en el proceso de análisis de inteligencia de la Comunidad de Inteligencia de los Estados Unidos (CI). Dentro del entorno de amenazas actual, aprovechar esta tecnología específica de Inteligencia Artificial (IA) es crucial para mantener la superioridad internacional. Se utilizará una metodología de análisis de contenido cualitativo para responder a la pregunta de investigación: ¿deberíamos introducir el PLN en el

1   Brandonbdm@gmail.com

doi: 10.18278/si.10.1.8

proceso de análisis de inteligencia? Este tipo de estudio permitirá la comparación de casos de uso del PLN en varios campos que realizan análisis de datos para identificar objetivamente cómo el PLN puede mejorar la práctica del análisis de inteligencia. La lente y el marco teórico elegidos para este estudio proporcionan su propia definición del proceso de análisis de inteligencia y consideran el PLN como un mecanismo para ampliar esa disciplina. El objetivo es llenar un vacío notable en la literatura existente y ofrecer valiosas perspectivas e implicaciones de la adopción del PLN dentro del proceso de análisis de inteligencia, considerando al mismo tiempo las limitaciones y sesgos asociados con los estudios cualitativos.

Palabras clave: Procesamiento del lenguaje natural; Comunidad de inteligencia de EE. UU.; Inteligencia artificial; análisis de inteligencia

情报分析过程中的自然语言处理

摘要

本研究旨在探究"将自然语言处理(NLP)融入美国情报界(IC)情报分析过程"一事的潜在优势与挑战。在当前的威胁环境下,利用这项人工智能(AI)技术对于保持国际优势至关重要。本研究将采用定性内容分析方法来回答以下研究问题:我们是否应该将NLP引入情报分析过程?此类研究将比较不同数据分析领域的NLP用例,以客观地识别NLP如何能改进情报分析实践。本研究选择的视角和理论框架对情报分析过程进行了独特的定义,并将NLP视为一种增强该学科的机制。本研究旨在填补现有文献中一个明显的空白,并在考虑定性研究的局限性和偏差的同时,提供一系列宝贵见解和启示,用于在情报分析过程中采用NLP。

关键词: 自然语言处理,美国情报界,人工智能,情报分析

Introduction

While the intelligence analysis process has provided the Intelligence Community (IC) a mechanism to supply decision-makers with intelligence critical to upholding national security, there are areas to improve upon. The enormous amount of data to analyze creates limitations in both processing and detecting patterns across a vast number of sources, as well as difficulties incorporating key variables from relevant sources. This study conducts research on and examines use cases of Natural Language Processing (NLP) applications across various fields to identify and present benefits that could significantly enhance the intelligence analysis process.

Strengthening the IC's analytical capabilities through NLP is critical in the current threat environment, which continually grows and adapts. NLP offers innovative methods for analyzing data and extracting relevant information from large datasets, emerging as a key advancement in intelligence analysis. NLP is bringing the IC the capability to help analysts rapidly transform ambiguous information into actionable intelligence. This technology doesn't merely serve as a tool to expedite analysis but has the potential to redefine it.

NLP not only introduces technological innovations but also provides intuitive methods to enhance intelligence analysis capabilities. While there is extensive research on how NLP can aid data analysis across various industries, there is a noticeable gap in the literature concerning its role in enhancing intelligence analysis. Therefore, it is essential to research applications, benefits, and challenges of NLP concerning its role in intelligence analysis.

The research question posed is: should we introduce NLP into the intelligence analysis process? The purpose of this study is to investigate mechanisms through which the IC, and the intelligence analysis conducted therein, could be improved. This will be achieved by conducting research and comparing the applications and outcomes of NLP used for data analysis across various industries, with the goal of identifying how NLP could enhance intelligence analysis. Integrating NLP tasks and applications has the potential to not only streamline current data analysis processes but also revolutionize the way information is transformed into intelligence and extracted from vast and diverse sources. The potential opportunities NLP offers the IC are explored along with the challenges associated with its integration. This research is pivotal for advancing the field of intelligence analysis within the current digital age and responding to evolving threats.

Literature Review

This literature review establishes the base of knowledge related to intelligence analysis that is conducted within the IC and NLP within the field of Artificial Intelligence (AI). The vast amount of data to be analyzed creates limitations in processing and detecting patterns across a large number of sources, as well as difficulties incorporating key variables from pertinent sources. A multitude of viewpoints, methods used in research, and prevailing arguments are explored. This section is divided into three portions: NLP, the intelligence analysis process, and the integration and compatibility of NLP and intelligence analysis.

Natural Language Processing

Artificial Intelligence

Szabadföldi (2021) explains that AI is a technology that supports existing functional applications and is "designed to solve specific problems, collecting, organizing, processing, analyzing, transmitting, and responding to larger data sets, suitable and capable of corresponding to the cognitive ability of the human intellect, and operations approaching it" (158). AI is reshaping the way humans address problems and transforms the methods used to carry out tasks that typically require cognitive processes, especially those associated with intelligence work. Ish, Ettinger, and Ferris (2021), who support the use of AI in intelligence analysis, mention AI is a computing technology that exhibits what humans would consider to be intelligent behavior. Its impact is significant enough that it has been listed as one of the emerging and disruptive technologies by NATO (Szabadföldi, 2021). NLP, while under the umbrella of AI, is a distinct type. Figure 1 provides a visual depicting the relationships between terms.

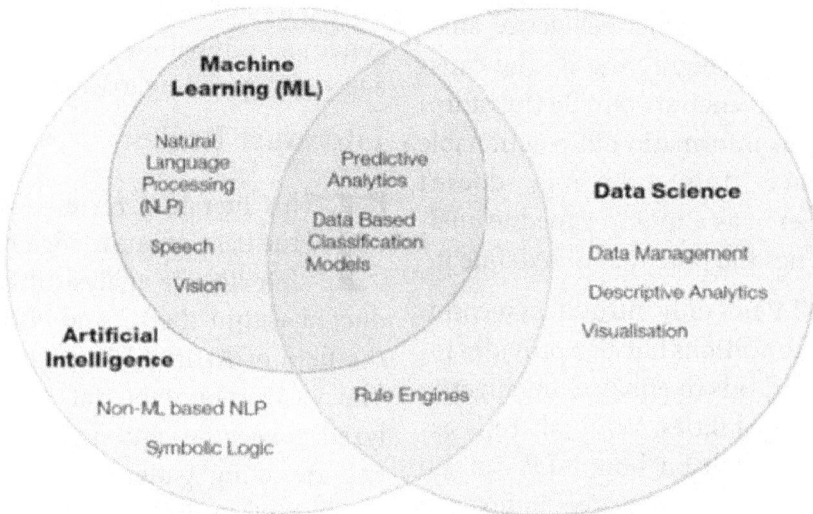

Figure 1. The Relationship Between AI, ML, and Data Science (Szabadföldi 2021)

Machine Learning

Machine Learning (ML) is a common term used interchangeably with AI. Although different, ML is a subset of AI. NLP is nested under and interrelated with ML (Cai, 2021). ML, with or without direct supervision, teaches computers with mathematical data models and can perform complex tasks with only limited human input (Szabadföldi, 2021; Shaffer and Shearn, 2023). NLP is created from the refinement and specialization of AI and ML technologies and tools.

Natural Language Processing

NLP is a complex term due to its diverse applications and is recognized as a field of research (Goel, 2017), an analysis technique, and a collective term referring to automatic computational processing of human language (Goldberg, 2017). The integration of NLP has been impactful across industries and recognized by many as a vital tool for data analysis (Crowston, Allen, and Heckman, 2012; Goel, 2017; Guetterman et al., 2018). NLP can use computational algorithms to learn, understand, and produce human language content, including the ability to accomplish useful and meaningful tasks (Guetterman et al., 2018; Cai, 2021). While there are various types of NLP, this study focuses on NLP operating within a neural network. A neural network is an AI model and method inspired by the structure and function of the human brain, designed to establish connections.

Neural networks offer powerful learning mechanisms, making them well-suited for natural language problems (Goldberg, 2017). NLP, grounded in ML, enables applications and tools to analyze, interpret, and generate text (Hariri, Fredericks, and Bowers, 2019). The definition of NLP for the context of this study is: the range of AI and ML capabilities that can analyze, understand, and produce human language including a variety of tools and applications in multiple forms. This definition offers a wide area for the findings of this study to apply NLP into the intelligence analysis process and allows organizations within the IC to refine the specific type or model of NLP that best fits their needs.

Intelligence Analysis Process

An analyst's expertise is essential in transforming raw data into actionable intelligence while also driving the intelligence process forward. Analysts perform functions along spectrums of

activities and are relied upon to make sense of partial and unclear data (Kreuzer, 2016). The mission of intelligence analysts within the IC is to evaluate, connect, apply context, infer meaning, and ultimately make analytic and operational judgments based on all available data including expectation of future threats (Office of the Director of National Intelligence, 2019). Kreuzer (2016) mentions factors that determine an analyst's performance are dependent on the time available to process and analyze data and the quantity of data and information available. Other critical parts to the equation include proper tradecraft, training, data, mindset, and cognitive ability. Kreuzer (2016) highlights that the intelligence analysis process is not a single activity that resides at one point in the intelligence process or setting—it's both art and science.

Intelligence analysis is defined in various ways, reflecting its diverse use across organizations. Rob Johnston's simple definition can be applied to many faucets of intelligence and does not confine the application to one discipline or area of intelligence. The definition states that intelligence analysis is a "socio-cognitive process, occurring with a secret domain, by which a collection of methods is used to reduce a complex issue to a set of simpler issues" (Ish, Ettinger, and Ferris, 2021, 27). Regens (2019) notes that solid analysis relies on applying the right techniques to clear, answerable questions to make sense of data. Kreuzer (2016) identifies the "unique expertise of intelligence analysis as the management of adversary information; consisting of the technical, methodological, knowledge, and organizational skills to facilitate a human organization whose primary function is the reduction of uncertainty in the decision-making process" (590). Reducing uncertainty is a theme echoed across the literature, as intelligence analysis is ultimately to prevent strategic surprise and identify new information.

Within intelligence analysis, data is conceived as raw, unprocessed material, and is the well-formed combination of meaningful data through processing and extraction. Intelligence is created through the combination and refinement of information to support decision-making (Blanchard and Taddeo, 2023). Whereas for Moran, Burton, and Christou (2023) analysis requires intuition, deliberative thinking, and using knowledge to turn raw data into predictions and policy options. These sources provide slightly divergent angles yet at the core offer an understanding of intelligence analysis as a vital tool in honing raw data into actionable intelligence and to facilitate sound decision-making. The definition of the intelligence analysis process to guide this study is the cognitive and technological supported process of comparing and contrasting various intelligence sources to identify key themes and indicators, incorporating new and relevant data for evaluation, making inferences, comprehending and conceptualizing data, detecting patterns, identifying and mitigating biases, and drawing conclusions. This definition is crafted from a culmination of the literature and is tailored for the purpose of this study.

Intelligence Analysis and Natural Language Processing

The above sections have provided a synthesis of relevant thoughts, themes, and definitions for intelligence analysis and NLP across the literature. NLP tasks are seen as integral for the DoD and IC due to their unique terminology laden with jargon, the sensitive nature of much of its data, and a constrained computing environment (Office of the Director of National Intelligence, 2019). NLP can analyze and comprehend both spoken and written human communication, marking a revolutionary stride in the intelligence practice and analysis deemed inevitable in the context of the national security functions which stand to benefit immensely from the adoption of NLP (Office of the Director of National Intelligence, 2019; Schirmer et al., 2021; Blanchard and Taddeo, 2023). This study comes from necessity to better understand the implications and effects NLP could have.

Field Specific Use Cases and Studies

Several studies sought NLP as a solution for news and media analysis. Kumar, Komalpreet, and Sukhpreet (2021) mention the difficulty in manually extracting useful information from text and through a survey paper provide an overview of work done in the field on an NLP driven solution. Listing several applications of NLP with examples of each, Ng et al. (2020) also describe common information extraction NLP uses from open-source news articles and media sources.

In the clinical field, NLP has become a heavy area for research. Text analytics through NLP have emerged as powerful tools in healthcare, revolutionizing patient care, clinical research, medication safety, and public health administration (Wong et al., 2018; Hasikin et al., 2023). This is because NLP facilitates the capturing of nuances of personal change and growth not currently captured by subjective symptom measures (Norman et al., 2020) and aids in difficult concept extraction that provides clear benefits (Wong et al., 2018). Also, it allows for information extraction from unstructured free-text notes that provide a wealth of information typically stored in electronic records to improve patient care, treatment, and decision-making (Kim et al., 2021; Yew et al., 2023). Chen et al. (2021) note that "applying NLP to text from other domains can provide answers to questions the literature does not yet address, for example, using social media and news to address public health issues" (4).

The evidence and analysis conducted within various data science applications provide additional terminology, ideas, and technologies to how NLP can be applied to intelligence analysis. NLP serves the ability to not only interpret human language, but to conduct analysis and provide insights on information across large datasets. Lawley et al. (2023) seek to fill a knowledge gap by integrating NLP with geological expertise to revolutionize geoscientific analyses. While Cai (2021) conducts a comprehensive review of the ways that researchers have utilized NLP in urban studies to develop a synthesis of opportunities and challenges for advancing urban research through the adoption of NLP. These studies present new ideas for data analysis methods while also providing additional insights to the range of data types, NLP can handle for the intelligence analysis process.

NLP studies are also expanding into fields that more closely resemble traditional intelligence analysis. The literature presents intelligence and law enforcement research into incorporating NLP to solve problems or enhance current analysis techniques. Sources describe NLP for analysis in crime related documents from foreign nations, redacted intelligence documents in Pakistan, national security and terrorism concerns, and data strategy for intelligence collection and analysis (Carnaz, Antunes, and Nogueira, 2021; Bridgelall, 2022; Mandrick and Smith, 2022; Shaffer and Shearn, 2023).

These sources and use cases will furnish evidence for this study. They demonstrate how NLP's detailed applications can translate effectively into the intelligence analysis process. Drawing from a rich pool of research allows for a variety of comparisons and scenarios NLP could be applied to. While these sources highlight the growing interest in utilizing NLP for intelligence analysis, they lack direct comparisons and insights to enhance the intelligence analysis process specifically, a gap this study aims to address.

Theoretical Framework

While the literature explores various

approaches to integrating NLP with intelligence analysis, Blanchard and Taddeo's (2023) AI-augmented intelligence analysis framework emerges as the most suitable for this study, emphasizing that NLP can assist but not replace the analyst. This study examines the research through the lens of AI-augmented intelligence analysis, focusing on how NLP can be integrated into the intelligence analysis process to enhance its capabilities, without replacing the human analyst. This framework views NLP as a tool to support intelligence analysis through cognitive automation. Cognitive automation "entails delegating to machines tasks which have been performed by humans thus far and which range across language processing, picking out patterns of speech, authorship attribution, classification and facial matching, and transcribing text from audio data for an analyst to search by using keywords or pre-set categories" (Blanchard and Taddeo, 2023, 6). This is reinforced by Regens (2019), who echoes that AI shall augment human cognition by leveraging high-performance computing.

Within the existing literature, there is a noticeable gap concerning how NLP can be effectively integrated into intelligence analysis processes. This study seeks to fill the gap by drawing comparisons with other fields that have successfully incorporated NLP into their data analysis techniques. Furthermore, the drawbacks and limitations of NLP will be discussed to provide a rounded and unbiased analysis of NLP's applicability for intelligence analysis. Based on the abilities of NLP

highlighted in the existing literature, NLP has the capability to shift the bottleneck in the intelligence process from the number of available analysts to the amount of data, a paramount issue in today's intelligence construct.

Methodology

This study aims to satisfy a recognized need for further research on how NLP can aid data analysis within the IC, using qualitative method with secondary and peer-reviewed sources. NLP was viewed as a method to augment intelligence analysis while examining interactions between NLP and human cognitive processes. The research question posed is: should we introduce NLP into the intelligence analysis process? The purpose is to research and compare the applications and outcomes of NLP across various use cases and research concerning NLP implemented in different industries to identify potential applications that could benefit the intelligence analysis process. This study is poised to contribute to meaningful advancements within the intelligence field and lay a foundational base for NLP in analysis for future studies.

Exploring the potential of NLP in intelligence analysis is necessary in order to showcase the best practices of producing actionable intelligence for decision-makers. Indeed, the impact of AI on the intelligence processes is not yet fully explored in the academic literature (Moran, Burton, and Christou, 2023). This is mainly due to the double wall of secrecy, which includes the sensitivity of intelligence sources and methods along with secrecy agreements involving tech collaborators who seek to protect their cutting-edge products (Moran, Burton, and Christou, 2023). The hypothesis to be supported or refuted is: use cases of NLP used for data analysis will qualitatively show benefits that can be applied within the intelligence analysis process.

Research Design

Qualitative research explores real-world problems by capturing experiences, perceptions, and behaviors, answering "how" and "why" questions, identifying patterns and themes, and tells a story (Tenny, J. Brannan, and G. Brannan, 2022). This study's approach involved qualitative text analysis to provide a rich and detailed understanding that is specific to NLP for augmenting the intelligence analysis process. This "... generally involves reading the data, assigning qualitative codes as succinct descriptors of meaning to text segments, and identifying themes that capture the major inferences to address study aims or research questions" (Guetterman et al., 2018, 2). The specific qualitative text analysis method this study uses is content analysis. The research for this study was conducted in several stages. First, data was collected, followed by an analysis that considers variables and procedures for bias mitigation. Finally, the research highlighted how NLP has benefited data analysis in various fields and draws comparisons to how the intelligence analysis process may benefit.

Data Collection

The data was collected mainly from secondary and peer-reviewed sources comprised of government publications, academic journals, technical documents in computer and data science, military handbooks, and industry reports. The selection of sources encompassed two distinct fields: NLP and intelligence analysis. Data collection used free text searches within academic libraries to access a diverse range of literature important to this study. The relevance of the selected data was refined through keyword searches. The keywords included: data analysis, text analysis, qualitative textual data analysis, unstructured data, structured data, semi-structured data, NLP, AI, intelligence analysis, process improvement, benefits, enhance, big data, and IC. Boolean operators AND and OR aids source selection, e.g., using "NLP AND Intelligence analysis" to amass relevant literature. Several NLP use cases and testbeds for data analysis across fields such as news, media, healthcare, education, and science are evaluated, formed the foundation of this study. Given the continual advancements in NLP and AI, the majority of the literature selected were those published within the last five years.

Data Analysis

Content analysis is a systematic and objective means of describing and quantifying phenomena that is non-intrusive (Elo et al., 2014). Qualitative content analysis has been defined as a research method for the subjective interpretation of the content of text data through the systematic classification process of coding and identifying themes or patterns (Hsieh and Shannon, 2005). There are three phases of content analysis. The preparation phase consists of collecting suitable data for content analysis, making sense of the data, and selecting the unit of analysis. Using an inductive approach, the organization phase involves open coding, creating categories, and abstraction, with themes, patterns, or codes emerging directly from the data in a data-driven, bottom-up process (Elo et al., 2014; Humble and Mozelius, 2022). The coding conducted enables the separation of different terms, themes, and concepts of the data for analysis. The reporting phase is conducted through the findings and analysis.

The coding is conducted by two means. One is color-coding different details from the data to highlight the differentiation. The other is coding of the data into major themes separated into different sections of a matrix. Rosen et al. (2023) mention that this process of charting data into a matrix, organizing information by rows and thematic codes in columns, is known as Framework Matrix Analysis, which is a highly structured approach for analyzing qualitative data. In the data, NLP themes such as applications, tools, effects, and goals achieved for the IC was used as codes. Analyzing the coded data revealed underlying themes and ideas that were not explicitly stated. The key concepts and findings derived from the coded matrix, generated through this research methodology, formed the primary sections of the findings and analysis. The framework of AI-augmented intelligence analysis served as a lens while conducting the content analysis.

Variables and Obstacles

This study is unique and presses the realm of measurable effects and is therefore intended to supply the comparisons of benefits from other fields as it provides relatable benefits to the intelligence analysis process. The independent variable is the incorporation of NLP into the intelligence analysis process. The dependent variable is the proposed benefits to the intelligence analysis process. For organizations within the IC that implement NLP within the intelligence analysis process, evaluating NLP effectiveness may involve exploring the perceived impact of NLP on task completion by comparing how human and NLP-driven performances are experienced and interpreted in the context of the task. Ish, Ettinger, and Ferris (2021) mention one way to measure performance is to compare the AI to what humans do. This is potential rating metric because "meeting or exceeding human performance is actually an especially strong criterion to enforce on the AI system" (Ish, Ettinger, and Ferris, 2021, 25). The operationalization to measure effects may be the feedback from customers using the produced intelligence.

Among the limitations of this study, the researcher does not possess the mechanism or resources to implement NLP into the IC to test the outcomes. Compounding this difficulty is the challenge of measuring success or effectiveness, given the varied analysis methods and diverse objectives across different IC organizations. Another limitation lies within the data collection process; this study does not capture unpublished, classified, or sensitive implementations of NLP in intelligence. A potential bias includes the favoritism for positive outcomes presented in the literature and could influence the interpretation of data. Therefore, all literature pertaining to the collection plan was analyzed regardless of NLP use case outcomes.

Findings and Analysis

The intelligence analysis process is used to provide actionable, timely, and relevant intelligence to decision-makers. This process provides abundant opportunities for enhancement, especially with emerging technologies such as NLP. The purpose of this study was to research and compare applications and outcomes of NLP across various use cases and industries to identify potential applications that could benefit the intelligence analysis process. Qualitative content analysis was performed using a theoretical framework based on AI-augmented intelligence analysis, as proposed by Blanchard and Taddeo (2023), who view AI as a tool to assist, rather than replace, analysts. The analysis was guided by the definition of the intelligence analysis process as a cognitive, technology-supported approach, involving theme identification, data evaluation, pattern recognition, bias mitigation, and inference drawing. Additionally,

the analysis incorporated the definition of NLP as the range of AI and ML techniques used to analyze, understand, and generate human language across various applications.

This section provides findings and analysis with the aim of supporting or refuting the hypothesis. Critical elements identified were NLP tasks and benefits. NLP tasks are uses of NLP either by technique, tool, or application to achieve a desired result. NLP benefits are defined as the observations noted as a result of implementing a set of or single NLP tasks. NLP tasks serve as the catalyst and are therefore presented in the order the coding revealed based on their frequency within the data.

NLP Tasks

This research found five main NLP tasks through content analysis. The tasks are listed in descending order based on their frequency identified from coding: text summarization, information extraction, information retrieval, question answering, and machine translation. NLP tasks involve the use of computers to process and analyze human language data, across all data structures, tailored to achieve specific objectives or desired outcomes (Wong et al., 2018; Lawley et al., 2023). Although there are more than five NLP tasks, this research identified these as most pertinent for the intelligence analysis process. For each of the following tasks a description of its function, application examples, comparison of uses across fields, and applicability in the intelligence analysis process are provided.

Text Summarization

The most common NLP task coding revealed was text summarization. Text summarization, or automatic summarization, conducts human-like language processing and produces a concise and understandable summary of a set of text (Crowston, Allen, and Heckman, 2012; Khurana et al., 2023). This process considers text length, syntax, and writing style in producing the output (Goel, 2017) while still preserving the important content which helps retain key information and makes summaries more efficient and effective (Kumar, Komalpreet, and Sukhpreet, 2021). Core content is prioritized during text summarization by considering sentence location, as important sentences tend to appear at the beginning or end of paragraphs, along with title similarity, term frequency, sentence position, and keyword retention (Kumar, Komalpreet, and Sukhpreet, 2021). These features are weighted and combined to preserve key information in the output. There are two different text summarization types. Etraction-based methods use key phrases directly from the source and abstraction-based methods, arguably the best, paraphrase for a more human-like summary (Ng et al., 2020; Kumar, Komalpreet, and Sukhpreet, 2021).

Evidence

There is a growing demand for summarization tools that not only capture essential information but also grasp deeper emotional meanings (Khurana et al., 2023). Kumar, Komalpreet, and Sukhpreet (2021) include the following benefits of summarization: reduces reading time, eases understanding through outlines, provides less biased algorithm-generated summaries, facilitates personalized data in question answering systems, expands the capacity of business intelligence services, supports multi-language summarization, enhances productivity without losing important details, and allows more information in smaller space. Goyal, Li, and Durrett's (2023) study on the potential of the GPT-3 large language model for summarization found that GPT-3 generates exceptional summaries that can be adapted to various settings. Challenges in summarization arise with the intricacies of human language and expressions, especially in transcribed texts that use adjectives, adverbs, and appositions. Another limitation identified is that these NLP tools are typically tailored for the English language (Goyal, Li, and Durrett 2023).

Implications for Intelligence Analysis

Summarization is important for the intelligence analysis process because analysts parse through large amounts of text to supply important information for analysis. Summarization serves as a tool to distill extensive or complex information into a concise and understandable format. This enables analysts to quickly grasp the content of documents without having to read the entirety making search results more actionable. Summarization is key for augmenting the intelligence analysis process by making the text more digestible for human readers (Chen et al., 2021). This task sifts through large amounts of data quickly, pulls relevant and critical terms, and provides the analyst with information faster than a human analyst, saving time. Blanchard and Taddeo (2023) echo this by iterating NLP, underpinned by neural networks, can now provide sophisticated mimicry of reading comprehension, summarization, "common sense" reasoning, provide simplicity in information queries, help organize data, and bypass many mundane tasks (Blanchard and Taddeo, 2023).

Information Extraction

Information extraction identifies phrases of interest from textual data and provides relevant information to a user with many data analysis applications (Ng et al., 2020; Khurana et al., 2023). Information extraction can identify various entities, including names, places, events, dates, times, and prices (Khurana et al., 2023). Information extraction transforms unstructured data into structured representations by recognizing and tagging key information. This task focuses on extracting semantic information pertaining to the meaning or interpretation derived from its original context (Goel, 2017) maintaining the most information validity.

Named Entity Recognition

Named Entity Recognition (NER) is a subset of information extraction (Kim et al., 2021; Otter, Medina, and Kalita, 2021; Zhou et al., 2022). The goal of NER is to identify, extract, and classify important named or specialized entities into predefined categories or classify them into different classes (Ng ct al., 2020; Kim et al., 2021; Khurana et al. 2023). NER is typically the initial step in downstream text processing tasks, aiming to provide semantic interpretations of the text by identifying and classifying specific concepts (Chen et al., 2021). NER is a versatile tool in data analysis due to its ability to accurately identify and categorize specific entities within large datasets, streamlining the process of information extraction and enabling more focused and contextual understandings.

Evidence

Information extraction and NER are used across various datasets and provide helpful insights into data. Yew et al. (2023) assert that conventional epidemiological research frequently depends on manual chart reviews, which are time-consuming, labor-intensive, and prone to errors. To address this, NLP information extraction has been employed for rapid and systematic automated extraction of patient characteristics from datasets. In the medical field, where managing vast volumes of documentation is paramount, information extraction and text summarization can facilitate data exploration and management by generating category labels, informative keywords, and concise summaries (Zhou et al., 2022).

Chen et al. (2021) examined a system designed for uncovering new inferences related to COVID-19 data. NER provided the essential data for knowledge discovery, streamlining a process that would have been considerably more time-consuming for researchers to undertake manually. In these domains, a significant portion of the data is unstructured. Kim et al. (2021) observed that free-text notes often hold more comprehensive information compared to structured data and information extraction techniques can process and present this richer information to analysts.

A challenge in NER involves selecting words in a specific context that are more informative, typically emphasizing noun phrases in text documents (Khurana et al., 2023). In the online environment slang or non-standard English can cause NLP tools to struggle processing the prediction of the writer's intent. To address this, models tailored to specific geographic areas could be developed and incorporate regional sarcasm, expressions, and informal phrases to mitigate these challenges (Khurana et al., 2023).

Implications for Intelligence Analysis

Information extraction reveals crucial details from various data types. This could provide analysts highlights of relevant and key information offering support for the conduct of intelligence analysis. Kim et al. (2021) highlight that extracting information from unstructured, free-text notes enhances patient care and decision-making. Similarly, in the realm of intelligence analysis, information extraction is crucial for delivering key intelligence. Additionally, this task's ability to predict future occurrences or effects, as seen in the medical field, represents a significant advancement in data analysis (Zhou et

al., 2022). Applying this predictive capability to the intelligence analysis process could lead to profound impacts, at the very least providing analysts new ideas and approaches to problem-solving. Such an approach amplifies the research's statistical power beyond what an analyst could achieve on their own (Yew et al., 2023).

Information Retrieval

Information retrieval aims to ensure individuals have access to the most relevant information, in the most suitable format, exactly when they require it. Filtering and indexing are used to as a set of techniques that enable users to quickly access relevant information from large text collections and identify pertinent subsets of documents within literature or databases (Goel, 2017; Chen et al., 2021; Otter, Medina, and Kalita, 2021). Information retrieval systems manage large document collections using three main types of algorithms. These algorithms include retrieval algorithms, which extract information either by scanning text or using an index; filtering algorithms, which simplify text to aid searches but may not distinguish between similar text fragments; and indexing algorithms, which create data structures for faster searches (Goel, 2017). Given the vast amount of data available, the robust capabilities of information retrieval tasks are crucial for organizing, processing, and delivering the right information.

Evidence

Leveraging a neural network for information retrieval allows the system to discern relationships within the data and identify word relationships (Otter, Medina, and Kalita, 2021). Yew et al. (2023) note NLP improves data analysis quality by reducing false negatives and identifies missing data through information retrieval. This task aids "analyses to be faster and use larger scales than manual analysis allows and has long been recognized as a way to alleviate information overload in biomedical research" (Chen et al., 2021, 3). It also showed accurate differentiation of epileptic seizures from psychogenic nonepileptic seizures and included simultaneous identification of multiple sudden unexplained death in epilepsy-relevant characteristics from electronic health records (Yew et al., 2023). The surge of unstructured textual information online and within medical databases has elevated the importance of information retrieval.

One challenge of information retrieval "pertains to ranking documents with respect to a query string in terms of relevance scores for ad hoc retrieval tasks, similar to what happens in a search engine" (Otter, Medina, and Kalita, 2021, 612). This highlights the need for these systems to effectively determine and score the relevance of documents to a specific search query. Another challenge involves the emergence of new trends or concepts. As Chen et al. (2021) discovered, NLP may have difficulties with information retrieval on terms and words it has never encountered.

Implications for Intelligence Analysis

Information retrieval serves to quickly provide users with pertinent data

from vast data collections. The findings concerning information retrieval qualitatively demonstrates similar uses and benefits for the intelligence analysis process. Information retrieval enables quicker access to knowledge, thereby increasing search efficiency (Zhou et al., 2022). This task is paramount to the intelligence analysis process as analysts require the most up to date and accurate data to make assessments. When new terms and concepts emerge, that data should be incorporated into information retrieval training algorithms or models to supply the best chance of relationship, source, and term identification (Chen et al., 2021).

A promising area for improvement lies in application programming interfaces (API) which enable different software applications to communicate. Chen et al. (2021) mention "in addition to interactive website access, most resources offer batch downloads and APIs for automated data retrieval" (12), establishing that information retrieval supported by APIs could provide additional avenues for information retrieval. Therefore, NLP tasks could be provided access to dispersed intelligence databases that typically require manual extraction enabling automated data retrieval (Chen et al., 2021) offering faster acquisition of actionable data from large datasets for analysis.

Question Answering

Chen et al. (2021) describe question answering as a combined task that takes a question and produces ranked or summarized answers. It combines techniques from information retrieval, text summarization, and literature-based discovery. This task has the ability to interpret and respond to human-posed questions by inferring and drawing answers from sources it is given permission to. While question answering shares similarities with summarization and information extraction, it distinguishes itself by extracting relevant words, phrases, or sentences from a document, then cohesively presenting this information in response to a query (Otter, Medina, and Kalita, 2021).

Evidence

Question answering empowers various industries to address challenges. Such systems enhance customer care services through real-time responses (Khurana et al., 2023) and can be integrated into chatbots to promptly address customer inquiries and provide solutions (Goel, 2017). NLP uses computational techniques to analyze and interpret human language. One particular deep learning model, Bidirectional Encoder Representations from Transformers (BERT), has revolutionized NLP by delivering precision comparable to human experts on benchmarks for question answering (Nassiri and Akhloufi, 2023). Khurana et al. (2023) indicate that recent advancements in question answering allow these tasks to truly "understand" queries, shifting from merely delivering search results to providing direct answers.

As the technology evolves, instead of presenting pages of data in response to a query, the system will recognize the query's deeper context, filter the data, and provide a clear and concise response rather than presenting raw data (Khurana et al., 2023). Schirmer et al. (2021) suggest question answering can be used to provide outputs for additional analysis. A concept proven when a "question answering tool was designed to help our own analysts build profiles of events and suspects, and the analysts needed to know specific facts from a case, such as whether the suspect knew his victims or had a prior arrest record" (Schirmer et al., 2021, 13). These systems not only elevate service quality and patient experience but also bolster clinical decision-making by extracting vital information such as family history, treatment data, and causal relations (Zhou et al., 2022). Some question answering NLP tasks integrate other tasks to create a comprehensive toolset for more accurate results.

Several challenges have been recognized for question answering. Limited training datasets pose difficulties in supplying effective models, and handling homonyms can be problematic for these NLP tasks where sarcasm or irony are often misinterpreted (Chen et al., 2021; Khurana et al., 2023). Ultimately, question answering has proven effective in equipping researchers and healthcare professionals with relevant information by comprehending queries and delivering impactful answers.

Implications for Intelligence Analysis

In intelligence analysis, question answering is invaluable for efficiently extracting answers from expansive datasets that would otherwise be tedious and time-consuming to find or interpret through analysts alone. The development of this task has helped clinical decision support by mirroring traditional patient-provider communication. Question answering facilitates data retrieval, enabling researchers, medical staff, and analysts to effortlessly tap into knowledge within datasets (Zhou et al., 2022). By supplying the necessary knowledge without the need for manual identification and retrieval, this task could alleviate the burden on analysts. Evidenced by the findings provided, question answering can efficiently extract specific details from expansive datasets, assisting intelligence analysts in discerning vital facts about events and individuals. This task enriches analysis by delivering in-depth insights from accessible sources providing additional avenues for sourcing evidence, data, and intelligence, ultimately crafting informed assessments faster for decision-makers.

Machine Translation

Goel (2017) names machine translation as one of the oldest NLP tasks that translates text between languages, without human intervention, by utilizing programs, dictionaries, and grammar. Machine translation uses statistical analysis of bilingual documents to determine typical translations of words or phrases based on context (Crowston, Allen, and Heckman, 2012). This language modeling process is important because the system generates several translations or transcription hypotheses and is an evolving NLP function (Goldberg, 2017).

Evidence

Current machine translation systems are either bilingual or multilingual and can be tailored to specific domains, enhancing accuracy by limiting potential substitutions (Goel, 2017). Machine translation "can be quite successful given a sufficiently large set of training examples and indeed has become the dominant mode of analysis for large corpora" (Crowston, Allen, and Heckman, 2012, 527). Zhou et al. (2022) explain that machine translation is crucial for current healthcare operations due to its ability to interpret various languages and its increasing accessibility. These efficient multilingual systems not only offer more accurate translations but save time compared to human translators (Zhou et al., 2022). One challenge for machine translation is the requirement for substantial training datasets (Goldberg, 2017). Larger training datasets may need to incorporate grammatical errors and cultural nuances to increase accuracy.

Implications for Intelligence Analysis

Machine translation serves to help make text readable (Chen et al., 2021) and, as Zhou et al. (2022) indicated, this NLP task could provide translation faster than human linguists or translators. Blanchard and Taddeo (2023) explain that this task can assist in processing data by translating foreign materials and generating text summaries, reducing the review time for intelligence analysts. The rapid translation of different languages enables faster production of actionable intelligence. This includes translation of signals, cyber, open source, or human intelligence, where text collected from sources or databases can be translated into the analyst's language for analysis.

In all, machine translation enhances intelligence analysis by quickly or automatically translating, transcribing, and interpreting languages. Conducting translations faster and more accurately than traditional human linguists enables a clearer understanding of threat communications, whether spoken, typed, or transmitted through other means. Clear understanding aids in situational awareness, intelligence collection, and counter-operations against threats. The following section analyzes the benefits of NLP when implemented as a comprehensive system. Integrating NLP into the intelligence analysis process creates optimal synergy between analysts and AI.

NLP Benefits

Across the five NLP tasks—text summarization, information extraction, information retrieval, question answering, and machine translation—both distinct and combined uses for augmenting intelligence analysis were discussed. Each task offers unique capabilities, serves various mission sets, and satisfies distinct intelligence analysis requirements. Despite their specific applications, NLP tasks provide an ability to conduct data analysis producing more insights faster or reduce the workload on human analysts by contributing towards optimizing the extraction, summary, retrieval, and communicative capabilities of textual data analysis.

Each task is designed to handle large amounts of data and engage requirements meaningfully. In real-world applications, these tasks do not have to work in silos, they can be designed to work in unison to complement each other. For example, information retrieval could supply the data necessary for a question answering query that is then shortened in text length by text summarization to produce a digestible response. Combining NLP tasks serves to bring the best versatility and accuracy to answer user requirements and needs. The utilization of these tasks, as presented by this study, showcases their importance for intelligence analysis especially in the growing big data environment.

NLP and Analysis

The Office of the Director of National Intelligence (2019) states that "leveraging artificial intelligence, automation, and augmentation technologies to amplify the effectiveness of our workforce will advance mission capability and enhance the IC's ability to provide needed data interpretation to decision-makers" (III). The IC was initially drawn to AI for its ability to help analysts manage "data smog," or the big data challenge. However, the IC is now expanding its focus to explore NLP's potential to enhance all stages of the intelligence process (Moran, Burton, and Christou, 2023). NLP emerges as a valuable tool by alleviating processing burden on human analysts and ensuring accurate, swift, and scalable results in big data analysis to discover information, hidden patterns, and unknown correlations in massive datasets (Hariri, Fredericks, and Bowers, 2019; Moran, Burton, and Christou, 2023).

NLP provides analysts the ability to discern associations between activities, individuals, and organizations across numerous documents and sources in reduced time, offering valuable insights into intelligence reports for research and analysis (Hariri, Fredericks, and Bowers, 2019; Shaffer and Shearn, 2023) while offsetting resource constraints (Guetterman et al., 2018). NLP can be used as an evaluation support system guided by the standards in Intelligence Community Directive (ICD) 203. With NLP in the intelligence analysis process loop, it helps mitigates the risk of investigator bias by supplying a different perspective aiding in objectivity (Ish, Ettinger, and Ferris, 2021; Yew et al., 2023). Incorporating measures to ensure objectivity supports ICD 203, which requires analysts within the IC to perform their analysis functions with objectivity and an awareness of their own assumptions and reasoning (Office of the Director of National Intelligence, 2023). Structured analytic techniques help analysts mitigate cognitive biases, manage information overload, and improve the rigor, consistency, and transparency of their thought processes (Chang et al., 2018). While upholding ICD 203, NLP further supports analysts by acting as a sounding board to facilitate these checks.

Various NLP tasks have demonstrated the ability to surpass human performance, marking NLP systems as one of the ways forward for intelligence (Ng et al., 2020) producing more depth in results than unassisted text analysis (Guetterman et al., 2018). "A study evaluating the time savings associated with an NLP system compared to manual review found that for every one hour of NLP system development, there was a time savings of twenty hours of manual review" (Wong et al., 2018, 828). NLP bolsters analytical efforts by coherently deciphering vast amounts of data, revealing hidden insights, and facilitating decision-making. The following section discusses NLP limitations along with potential mitigation strategies to offset them.

Challenges and Limitations

There are several challenges the data revealed about using NLP to augment the intelligence analysis process. Current NLP systems offer impressive analytical capabilities that surpass existing human process efficiencies. However, this doesn't negate or resolve all issues, nor does this research deny the presence of certain limitations. Along with these challenges, this research proposes mitigating practices. Like all AI systems, NLP must be evaluated and monitored to ensure optimal accuracy and performance, while remaining free of biases and operating within ethical guidelines.

Evaluation

In the rapidly advancing context of NLP, defining effectiveness within the realm of intelligence analysis remains a complex effort. "Unfortunately, when analyzing such systems in the context of intelligence, effectiveness is not so obviously defined and measured; the procedure ODNI describes measures whether consumers are satisfied, not what impact the reporting generates" (Ish, Ettinger, and Ferris, 2021, 20). One way to evaluate the performance of an NLP system for aiding intelligence analysis is to hold it to the same standards as a human analyst after structured analytic techniques have been incorporated and applied (Ish, Ettinger, and Ferris, 2021). Accurate assessment of these systems hinges on establishing and continuously reevaluating appropriate metrics, maintaining clear communication, and conducting further research.

Tools and Development

Within intelligence organizations, technical expertise may pose a limitation in developing specialized NLP systems. The Augmenting Intelligence using Machines (AIM) Strategy "provides the framework for the incorporation of AIM technologies to accelerate mission capability development across the IC" (Office of the Director of National Intelligence, 2019, III). The AIM strategy supports objectives that promote asset sharing and AI security, enabling the enhancement of intelligence capabilities through collaboration. NLP models could be collaboratively developed with implementation plans, contingencies for potential failures, and proactive measures to counter foreign AI capabilities, collectively safeguarding these shared technological investments. Developing NLP tools not only hinges on technological advancements but on a collective and interdisciplinary endeavor to align them with requirements

of intelligence analysis (Schirmer et al., 2021).

Dispersed Data

"A key barrier to the utility of NLP is that of interoperability between data sources" (Wong et al., 2018, 838). Addressing the siloing of intelligence information within the IC is an area where NLP could be beneficial, especially if more organizations share and grant these tools access to information. Intelligence analysts often expend substantial time gathering and organizing data, navigating through abundant information. Many systems do not provide an API that enables automatic data pulls (Schirmer et al., 2021). Establishing APIs with NLP facilitates access and data retrieval from dispersed data sources. Leveraging APIs to enhance data interoperability could significantly streamline intelligence analysis, minimize manual data handling, and foster a more integrated approach across various IC platforms and organizations.

Black Box

A black box refers to AI systems, particularly those using deep learning with numerous layers in a neural network, that are not transparent or understandable to users; challenging interpretation or validation of the system's decision-making (Szabadföldi, 2021). This produces challenges with transparency and reproducibility of conclusions where NLP approaches may appear less appealing (Yew et al., 2023). Given that the most capable and high-performing NLP systems function within neural networks (Goldberg, 2017; Zhou et al.,

2022; Blanchard and Taddeo, 2023), the black box issue may arise for some organizations. Since this type of AI cannot reveal how conclusions are derived, it may not be suitable for high-level decision-making. This is another reason why this study utilizes an AI-augmented intelligence analysis framework.

Conclusion

This study set out to answer the research question: should we introduce NLP into the intelligence analysis process? By using a framework that treats NLP as an AI-augmented method for intelligence analysis, this study qualitatively identifies benefits relevant to the intelligence analysis process, supporting the hypothesis that use cases of NLP in data analysis will demonstrate benefits applicable within the intelligence analysis process. Additionally, challenges and limitations with respective mitigating solutions were provided.

Tasks and Benefits

NLP is a force multiplier that provides solutions to combat the growing amount of information that needs to be analyzed within the IC. Text summarization enhances the intelligence analysis process by cutting down reading time (Kumar, Komalpreet, and Sukhpreet, 2021), parsing through extensive text, and rendering content more accessible for readers (Chen et al., 2021). By producing concise text outputs, it allows analysts to concentrate more on pivotal data interpretations for decision-making, significantly enriching the intelligence analysis workflow.

Information extraction improves data utilization efficiency by highlighting pertinent and essential details from vast data sources (Zhou et al., 2022). It uncovers frequently missed events, enhancing the research's statistical robustness (Yew et al., 2023). Moreover, it reduces biases and boosts the quality of analysis (Zhou et al., 2022). Information retrieval facilitates quicker access to vital knowledge, enhancing search efficiency (Zhou et al., 2022). This task expedites the acquisition of actionable data from extensive datasets, aiding analysts in finding the correct information (Chen et al., 2021).

Question answering systems alleviate burden by providing insights from queries within data and offer curated responses that counter misinformation (Chen et al., 2021). Question answering efficiently extract answers replacing the time-consuming manual retrieval process and provide easy access into essential knowledge within datasets (Zhou et al., 2022). Integrating question answering streamlines evidence and data sourcing, leading to quicker and better-informed decisions.

Machine translation makes different languages readable for analysis in near-real-time (Chen et al., 2021). It provides translation faster than human linguists or translators (Zhou et al., 2022) and creates text summaries reducing the review time for intelligence analysts (Blanchard and Taddeo, 2023). These NLP tasks have proven to be beneficial across an array of industries and research fields.

NLP is important for superiority over adversaries. This technology offers a diverse range of capabilities that reveal insights and hidden patterns to facilitate informed decision-making. It not only enhances analytical processes, addressing the "data smog" issue (Moran, Burton, and Christou, 2023), but also aids in deriving accurate intelligence from complex datasets (Yew et al., 2023). Given the escalating threat environment with continuously growing data, NLP stands as the best mechanism to address intelligence analysis pitfalls.

Challenges

As a continually developing technology, NLP has areas for improvement. Along with ensuring NLP is functioning optimally, evaluating an NLP system's performance is complex necessitating properly chosen and continuously revised metrics. Human and NLP collaborative evaluation is important for effective intelligence assessments. Though developing NLP tools requires technical expertise, the AIM Strategy fosters interdisciplinary collaborations between computer scientists and domain specialists (Office of the Director of National Intelligence, 2019). Organizations together can enhance tool development through sharing of tools and the manpower to facilitate their production.

Dispersed data within the IC can challenge NLP tasks due to the siloed information. By combining NLP with APIs, data interoperability is enhanced, intelligence analysis is streamlined, and manual data handling is reduced, promoting collaboration across different IC platforms (Wong et al., 2018; Schirmer et al., 2021). The black box dilemma may complicate trust of officials with NLP operating within neural networks. These systems do not provide or track their decision-making process, making them less appealing (Yew et al., 2023) for decision-making despite their proven effectiveness in data analysis.

Contributions to Knowledge

This study researched and compared the applications and outcomes of NLP across various fields and use cases. It demonstrated the benefits of NLP in its applications and qualitatively compared data to show how it enhances the intelligence analysis process with associated advantages. The research aimed to fill a gap in the literature concerning the utilization of NLP for intelligence analysis.

NLP plays a crucial role in driving technological advancements that shape the future of intelligence analysis within the IC. NLP addresses the challenges presented by big data, reduced timelines, and increasing complex threats. It leads to improved efficiency, decreased human error, and enhanced insights into potential threats and opportunities for faster intelligence analysis. Future research into how threats could attempt to poison or inject false data into neural networks and NLP models may yield methods to counteract such attempts. Additionally, further studies on ethics may help inform future decisions regarding the implementation of NLP in various areas within the IC.

Brandon Morad holds an MA in Intelligence Studies. His research focuses on the integration of Natural Language Processing within the U.S. Intelligence Community for intelligence analysis. He examines the evolution, structures, and functions of intelligence analysis and explores various intelligence collection methods. His work also evaluates traditional and emerging threats to national security. He welcomes opportunities for collaboration and further research.

References

Blanchard, Alexander and Mariarosaria Taddeo. (2023). "The Ethics of Artificial Intelligence for Intelligence Analysis: A Review of the Key Challenges with Recommendations." *Digital Society* 2, 12. https://doi.org/10.1007/s44206-023-00036-4.

Bridgelall, Raj. (2022). "An Application of Natural Language Processing to Classify what Terrorists Say they Want." *Social Sciences* 11 (1): 23. https://doi.org/10.3390/socsci11010023.

Cai, Meng. (2021). "Natural Language Processing for Urban Research: A Systematic Review." *Heliyon* 7 (3): e06322–e06322. https://doi.org/10.1016/j.heliyon.2021.e06322.

Carnaz, Gonçalo, Mário Antunes, and Vitor Beires Nogueira. (2021). "An Annotated Corpus of Crime-Related Portuguese Documents for NLP and Machine Learning Processing." *Data (Basel)* 6 (7): 71–. https://doi.org/10.3390/data6070071

Chang, Welton, Elissabeth Berdini, David R. Mandel, and Philip E. Tetlock. (2018). "Restructuring Structured Analytic Techniques in Intelligence." *Intelligence & National Security* 33 (3): 337–56. https://doi.org/10.1080/02684527.2017.1400230

Chen, Qingyu, Robert Leaman, Alexis Allot, Ling Luo, Chih-Hsuan Wei, Shankai Yan, and Zhiyong Lu. (2021). "Artificial Intelligence (AI) in Action: Addressing the COVID-19 Pandemic with Natural Language Processing (NLP)." *arXiv.org*. https://doi.org/10.1146/annurev-biodatasci-021821-061045.

Crowston, Kevin, Eileen E. Allen, and Robert Heckman. (2012). "Using Natural Language Processing Technology for Qualitative Data Analysis." *International Journal of Social Research Methodology* 15 (6): 523–43. https://doi.org/10.1080/13645579.2011.625764.

Elo, Satu, Maria Kääriäinen, Outi Kanste, Tarja Pölkki, Kati Utriainen, and Helvi Kyngäs. (2014). "Qualitative Content Analysis: A Focus on Trustworthiness." *SAGE*

Open 4 (1): 215824401452263–. https://doi.org/10.1177/2158244014522633.

Goel, Bhargavi. (2017). "Developments in the Field of Natural Language Processing." *International Journal of Advanced Research in Computer Science* 8 (3) (03). ISSN No. 0976-5697. http://www.ijarcs.info/index.php/Ijarcs/article/view/2944.

Goldberg, Yoav. (2017). *Neural Network Methods for Natural Language Processing. Synthesis Lectures on Human Language Technologies #37.* Springer Nature. doi:10.2200/S00762ED1V01Y201703HLT037.

Goyal, Tanya, Junyi Jessy Li, and Greg Durrett. (2023). "News Summarization and Evaluation in the Era of GPT-3." *Computer Science, Computation and Language.* Department of Computer Science, Department of Linguistics. The University of Texas at Austin. https://arxiv.org/pdf/2209.12356.pdf.

Guetterman, Timothy C., Tammy Chang, Melissa DeJonckheere, Tanmay Basu, Elizabeth Scruggs, and Vinod Vydiswaran V.G. (2018). "Augmenting Qualitative Text Analysis with Natural Language Processing: Methodological Study." *Journal of Medical Internet Research* 20 (6) (06). https://doi.org/10.2196/jmir.9702.

Hariri, Reihaneh H., Erik M. Fredericks, and Kate M. Bowers. (2019). "Uncertainty in Big Data Analytics: Survey, Opportunities, and Challenges." *Journal of Big Data* 6 (1): 1–16. https://doi.org/10.1186/s40537-019-0206-3.

Hasikin, Khairunnisa, Khin Wee Lai, Suresh Chandra Satapathy, Kadir Sabanci and Muhammet Fatih Aslan. (2023). "Editorial: Emerging Applications of Text Analytics and Natural Language Processing in Healthcare." *Front Digit Health.* 5:1227948. doi:10.3389/fdgth.2023.1227948.

Hsieh, Hsiu-Fang and Sarah E. Shannon. (2005). "Three Approaches to Qualitative Content Analysis." *Qualitative Health Research*, 15 (9), 1277–1288. https://doi.org/10.1177/1049732305276687.

Humble, Niklas and Peter Mozelius. (2022). "Content Analysis or Thematic Analysis: Doctoral Students' Perceptions of Similarities and Differences: EJBRM." *Electronic Journal of Business Research Methods* 20 (3): 89–98. https://doi.org/10.34190/ejbrm.20.3.2920.

Ish, Daniel, Jared Ettinger, and Christopher Ferris. (2021). *Evaluating the Effectiveness of Artificial Intelligence Systems in Intelligence Analysis.* Santa Monica, CA: RAND Corporation. https://www.rand.org/pubs/research_reports/RRA464-1.html.

Khurana, Diksha, Aditya Koli, Kiran Khatter, and Sukhdev Singh. (2023). "Natural

Language Processing: State of the Art, Current Trends and Challenges." *Multimedia Tools and Applications* 82 (3) (01): 3713–3744. https://doi.org/10.1007/s11042-022-13428-4.

Kim, Youngjun, Paul M. Heider, Isabel R. H. Lally, and Stéphane M. Meystre. (2021). "A Hybrid Model for Family History Information Identification and Relation Extraction: Development and Evaluation of an End-to-End Information Extraction System." *JMIR Medical Informatics* 9 (4) (04). https://doi.org/10.2196/22797.

Kreuzer, Michael P. (2016). "Professionalizing Intelligence Analysis: An Expertise and Responsibility Centered Approach." *Intelligence and National Security* 31 (4): 579–97. https://doi.org/10.1080/02684527.2015.1039228.

Kumar, Yogesh, Kaur Komalpreet, and Kaur Sukhpreet. (2021). "Study of Automatic Text Summarization Approaches in Different Languages." *The Artificial Intelligence Review* 54 (8) (12): 5897–5929. https://doi.org/10.1007/s10462-021-09964-4.

Lawley, Christopher J. M., Michael G. Gadd, Mohammad Parsa, Graham W. Lederer, Garth E. Graham, and Arianne Ford. (2023). "Applications of Natural Language Processing to Geoscience Text Data and Prospectivity Modeling." *Natural Resources Research (New York, N.Y.)* 32 (4): 1503–27. https://doi.org/10.1007/s11053-023-10216-1.

Mandrick, Bill, and Barry Smith. (2022). "Philosophical Foundations of Intelligence Collection and Analysis: A Defense of Ontological Realism." *Intelligence and National Security* 37 (6): 809–19. https://doi.org/10.1080/02684527.2022.2076330.

Moran, Christopher, Joe Burton, and George Christou. (2023). "The US Intelligence Community, Global Security, and AI: From Secret Intelligence to Smart Spying." *Journal of Global Security Studies,* Volume 8, Issue 2, June 2023, ogad005, https://doi.org/10.1093/jogss/ogad005.

Nassiri, Khalid, and Moulay Akhloufi. (2023). "Transformer Models Used for Text-Based Question Answering Systems." *Applied Intelligence (Dordrecht, Netherlands)* 53 (9): 10602–35. https://doi.org/10.1007/s10489-022-04052-8.

Ng, Victoria, Erin E Rees, Jingcheng Niu, Abdelhamid Zaghool, Homeira Ghiasbeglou, and Adrian Verster. (2020). "Application of Natural Language Processing Algorithms for Extracting Information from News Articles in Event-Based Surveillance." *Canada Communicable Disease Report* 46 (6): 186–91. https://doi.org/10.14745/ccdr.v46i06a06.

Norman, Kim P., Anita Govindjee, Seth R. Norman, Michael Godoy, Kimberlie L. Cerrone, Dustin W. Kieschnick, and William Kassler. (2020). "Natural Language Processing Tools for Assessing Progress and Outcome of Two Veteran Populations: Cohort Study from a Novel Online Intervention for Posttraumatic Growth." *JMIR Formative Research* 4 (9) (09). https://doi.org/10.2196/17424.

Office of the Director of National Intelligence. (2019). *The AIM Initiative: A Strategy for Augmenting Intelligence Using Machines.* https://www.dni.gov/files/ODNI/documents/AIM-Strategy.pdf.

Office of the Director of National Intelligence. (2023). *Analytic Standards: ICD 203.* https://www.dni.gov/files/documents/ICD/ICD-203.pdf

Otter, Daniel, Julian Medina, and Jugal Kalita. (2021). "A Survey of the Usages of Deep Learning for Natural Language Processing." *IEEE Transaction on Neural Networks and Learning Systems* 32 (2): 604–24. https://doi.org/10.1109/TNNLS.2020.2979670.

Regens, James L. (2019). "Augmenting Human Cognition to Enhance Strategic, Operational, and Tactical Intelligence." *Intelligence & National Security* 34 (5) (08): 673-687. https://doi.org/10.1080/02684527.2019.1579410.

Rosen, Rochelle K., Monique Gainey, Sabiha Nasrin, Stephanie C. Garbern, Ryan Lantini, Nour Elshabassi, Sufia Sultana, et al. (2023). "Use of Framework Matrix and Thematic Coding Methods in Qualitative Analysis for mHealth: NIRU-DAK Study Data." *International Journal of Qualitative Methods* 22. https://doi.org/10.1177/16094069231184123.

Schirmer, Peter, Amber Jaycocks, Sean Mann, William Marcellino, Luke J. Matthews, John David Parsons, and David Schulker. (2021). *Natural Language Processing: Security- and Defense-Related Lessons Learned.* Santa Monica, CA: RAND Corporation. https://www.rand.org/pubs/perspectives/PEA926-1.html.

Shaffer, Ryan and Benjamin Shearn. (2023). Advancing Intelligence Analysis: Using Natural Language Processing on East Pakistani Intelligence Documents. *Intelligence and National Security*, 38:5, 740–763, doi:10.1080/02684527.2023.2170744.

Szabadföldi, István. (2021). "Artificial Intelligence in Military Application – Opportunities and Challenges." *Land Forces Academy Review* 26 (2): 157–165. https://doi.org/10.2478/raft-2021-0022.

Tenny, Steven, Janelle Brannan, and Grace Brannan. (2022). "Qualitative Study."

StatPearls (Treasure Island, F.L.) https://www.ncbi.nlm.nih.gov/books/NBK4703 95/

Wong, Adrian, Joseph M. Plasek, Steven P. Montecalvo, and Li Zhou. (2018). "Natural Language Processing and Its Implications for the Future of Medication Safety: A Narrative Review of Recent Advances and Challenges." *Pharmacotherapy* 38 (8): 822–41. https://doi.org/10.1002/phar.2151.

Yew, Arister N. J., Marijn Schraagen, Willem M. Otte, and Eric Diessen. (2023). "Transforming Epilepsy Research: A Systematic Review on Natural Language Processing Applications." *Epilepsia* (Copenhagen) 64 (2): 292–305. https://doi.org/10.1111/epi.17474.

Zhou, Binggui, Guanghua Yang, Zheng Shi, and Shaodan Ma. (2022). "Natural Language Processing for Smart Healthcare." *IEEE Reviews in Biomedical Engineering* PP: 1–17. https://doi.org/10.1109/RBME.2022.3210270.

Active Shooter Awareness and Preparednessin Soft Target Scenarios

Joshua E. Lane[1]

Abstract

The incidence of active shooter scenarios continues to increase in the United States, with a 53% increase in 2021 (Federal Bureau of Investigation [FBI], 2022). The FBI defines an active shooter as "one or more individuals actively engaged in killing or attempting to kill people in a populated area" (FBI, 2022). The term active shooter defines the weapon utilized as a firearm (FBI, 2022). In 2021, a total of 61 active shooter incidents results in 103 fatalities and 140 wounded (FBI, 2022). Active shooter scenarios typically occur at soft target locations, that is one that does not have hardened security and/or defenses. These include schools and typical work locations. While the term active shooter has become commonplace in America, awareness and preparedness are not necessarily present. This study aims to explore active shooter awareness and preparedness. Specifically, a survey of individuals as well as an exhaustive review of the literature will be employed to provide both a quantitative and qualitative analysis. Awareness and preparedness are broad topics that carry vastly different meanings among various individuals.

Thus, realistic approaches must be considered in the education of these topics. Awareness and preparedness are core tenets of mitigation for active shooter situations.

Keywords: active shooter; soft target; awareness; preparedness; terrorism

Conciencia y preparación ante tiradores activos en escenarios de blancos fáciles

Resumen

La incidencia de tiradores activos continúa aumentando en Estados Unidos, con un incremento del 53% en 2021 (Oficina Federal de Investigaciones [FBI], 2022). El FBI define a un tirador activo como "una o más personas que participan activamente en matar o intentar matar personas en una zona poblada" (FBI, 2022). El término

1   jlane@lanederm.com

doi: 10.18278/si.10.1.9

tirador activo define el arma utilizada como arma de fuego (FBI, 2022). En 2021, un total de 61 incidentes con tiradores activos resultaron en 103 muertes y 140 heridos (FBI, 2022). Los escenarios de tiradores activos generalmente ocurren en blancos fáciles, es decir, donde no se cuenta con seguridad reforzada. Estos incluyen escuelas y lugares de trabajo típicos. Si bien el término tirador activo se ha vuelto común en Estados Unidos, la concientización y la preparación no necesariamente están presentes. Este estudio tiene como objetivo explorar la concientización y la preparación ante tiradores activos. Específicamente, se empleará una encuesta individual, así como una revisión exhaustiva de la literatura, para proporcionar un análisis tanto cuantitativo como cualitativo. La concienciación y la preparación son temas amplios que conllevan significados muy diferentes según el individuo. Por lo tanto, se deben considerar enfoques realistas en la educación sobre estos temas.

La concienciación y la preparación son principios fundamentales para la mitigación en situaciones de tiradores activos.

Palabras clave: tirador activo; objetivo fácil; concienciación; preparación; terrorismo

软目标场景中的现场行凶枪手意识和准备

摘要

在美国，现场行凶枪手(active shooter)事件的发生率持续上升，2021年增长了53%(FBI, 2022)。联邦调查局(FBI)将现场行凶枪手定义为"在人口稠密地区积极参与杀害或企图杀害他人的一名或多名个人"(FBI, 2022)。"现场行凶枪手"使用枪械武器(FBI, 2022)。2021年，共发生61起现场行凶枪手案，造成103人死亡，140人受伤(FBI, 2022)。现场行凶枪手案通常发生在软目标区域，即没有强化安保和/或防御措施的区域。这些区域包括学校和典型的工作场所。虽然现场行凶枪手一词在美国已变得司空见惯，但人们的意识和准备程度并不一定存在。本研究旨在探讨人们对现场行凶枪手的意识和准备程度。具体而言，使用了一项个体调查和详尽的文献综述来进行定量和定性分析。意识和准备是一个广泛的主题，在不同个体中含义迥异。因此，在开展这些主题教育时，必须考虑切实可行的方法。

意识和准备是应对现场行凶枪手事件的核心原则。

关键词：现场行凶枪手，软目标，意识，准备，恐怖主义

An active shooter is defined as "an individual actively engaged in killing or attempting to kill people in a populated area" (Federal Bureau of Investigation [FBI], 2021). The occurrence of such acts of violence is particularly frequent at "soft targets" (U.S. Department of Homeland Security, 2019). Soft targets are defined as "locations that are easily accessible to large numbers of people and that have limited security or protective measures in place making them vulnerable to attack" (U.S. Department of Homeland Security, 2018). The FBI (2022) reported a total of 61 active shooter incidents in the United States in 2021, representing a 53% increase from 2020.

The presence and increasing incidence of active shooter scenarios in the United States (and globally) is a growing threat to Americans. This is particularly a threat to children in the school setting.

Methods

The current study utilized a mixed method (quantitative and qualitative) research design to analyze the hypothesis. The purpose of the study was to examine current characteristics and protocols designed to mitigate active shooter situations in soft target scenarios. The study combined a quantitative survey assessing individual active shooter awareness and preparedness and a qualitative review of the literature.

Patients visiting a medical facility were offered the opportunity to participate in this voluntary survey. A consent was obtained and then the survey was provided to the individual. Participants were notified of the voluntary nature of the survey and that they may stop at any point if so desired. Completion of the survey took less than one minute typically and consisted of checking an answer to twelve brief questions. No minors or individuals of specialized populations were allowed to participate in this study. The survey (Figure 1) consisted of 12 questions related to individual awareness and preparedness regarding an active shooter scenario. Two of the questions asked the participants' age and gender. The answers to these questions are analyzed in the results section and aim to provide a snapshot of the everyday preparedness and perception of an individual's mindset regarding an active shooter situation. This study was approved by a local institutional review board.

The quantitative (survey) component of the current study was analyzed statistically. Microsoft Excel (version 2211) and SAS analytical software were used to enter and analyze the survey data.

Results

A total of 85 participants (64 % women; 35 % men) participated in the voluntary survey between October 2022 and November 2022. The average age was 44.8 (SD = 16.5), with a range between 19 and 84. The majority of survey participants worked in the healthcare field (46.9%), with education, financial, and retail following. The overwhelming majority (97.6%) of survey participants were familiar with the term "active shooter." However, 64%

ACTIVE SHOOTER AWARENESS & PREPAREDNESS SURVEY

1. What is your age? _____

2. What is your gender? ☐ M ☐ F

3. Are you familiar with the term "active shooter?" ☐ Yes ☐ No

4. What type of office do you work in?
   ☐ Education
   ☐ Healthcare
   ☐ Retail
   ☐ Financial
   ☐ Other: _____

5. Rank the importance of lockdown drills and other safety procedures? _____
   (Scale of 1 to 10, with 10 being most)

6. Do you believe that "active shooter" training is useful for a place of employment? ☐ Yes ☐ No

7. Do you currently undergo "active shooter" training at your place of employment? ☐ Yes ☐ No

8. How well do you think you are prepared for an "active shooter" incident? _____
   (Scale of 1 to 10, with 10 being most)

9. How well is your place of employment is prepared for an "active shooter" incident? _____
   (Scale of 1 to 10, with 10 being most)

10. What are your biggest challenges/concerns in the even of an "active shooter" incident? (if more
    than 1 please rank in order of importance [i.e. 1 being the most important]):

    ☐ personal safety and well-being (both yourself and others around you)
    ☐ communicating to people involved in the situation
    ☐ locating people involved in the situation
    ☐ making decisions during such a situation
    ☐ providing real-time updates to those involved
    ☐ coordinating with law enforcement

11. Does your place of employment have an "active shooter" plan in place? ☐ Yes ☐ No

12. If yes to Question 11, do you think this plan seems adequate? ☐ Yes ☐ No

Figure 1. Survey deployed to participants.

of participants stated that their place of employment did not have an active shooter plan. Among participants who stated that their place of employment did have an active shooter plan in place, 86% indicated that this plan seemed adequate. A majority (96.5%) further noted that active shooter training in the workplace is indeed useful; however, 74% of participants did not have any actual active shooter training at their workplace. Survey of individual perceived greatest challenges and/or concerns in the active shooter situation revealed that personal safety ranks among the highest (94.1%) (Figure 2).

Primary Challenge/Concern in an Active Shooter Incident

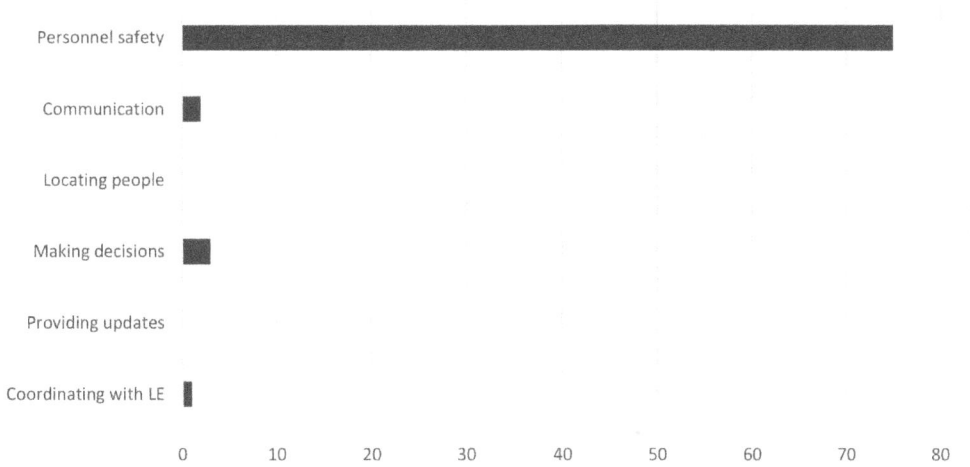

Figure 2. Primary challenge/concern of survey participants in an active shooter incident.

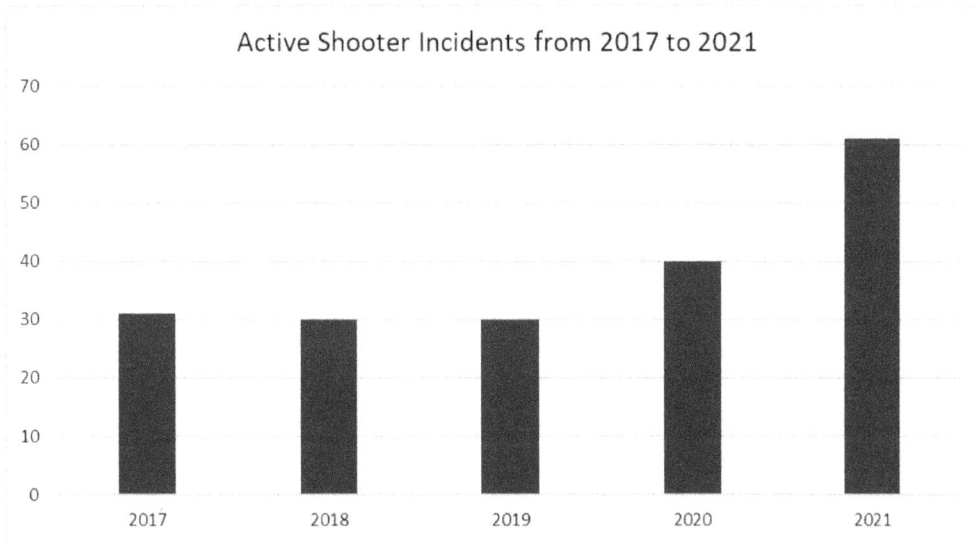

Figure 3. Number of active shooter incidents in the United States from 2017 to 2021.

Discussion

The FBI categorized a total of 61 shootings in the year 2021 as active shooter incidents. This represents an approximately 53% increase between 2020 and 2021 (Figure 3) (U.S. Department of Justice, 2021).

The FBI identified a total of 160 active shooter incidents in the U.S. between 2000 and 2013 (U.S. Department of Justice, 2018). Data is available for each of these years reported; however, the data for 2021 indicated 243 casualties, with 103 fatalities and 140 wounded. However, another study reported 154 shootings in U.S. hospitals between 2000 and 2011, thus indicating the inaccuracy in tracking such incidents (Kelen et al., 2012). Yet another study found 88 shootings in U.S. hospitals from 2012 to 2016 (Wax et al., 2019). Both studies analyzing shootings in hospitals demonstrated the emergency department as the most common site in the hospital (Kelen, 2012; Wax, 2019). The definition of active shooter that is used by U.S. government agencies is "an individual actively engaged in killing or attempting to kill people in a confined and populated area" (U.S. Department of Justice, 2018).

The FBI discovered that active shooter incidents occurred in both small and large towns and in almost all (40 of 50) states. They were most frequent in either commerce or educational environments (U.S. Department of Justice, 2018). Perhaps the most important finding was the fact that damage often occurs within minutes and is often of short duration. Approximately 70% of active shooter incidents end within 5 minutes, with 37% ending within 2 minutes (U.S. Department of Justice, 2018). This emphasizes the unnerving reality that law enforcement cannot react quickly enough in a majority of situations to ensure public safety. The amount of mortality and morbidity that can occur within minutes is tremendous. However, there are a multitude of actions that individuals present in an active shooter scenario may be able to do in an effort to improve survival.

Preparation for an active shooter incident is difficult as there are too many variables to definitively consider in entirety. These factors include characteristics about the perpetrator, the victims, the surrounding environment, the time of day, the ability of the offender, the psychological state of the offender, law enforcement response time, and many other factors. Response to such a situation may be comparable to an individual responding to a medical code situation. While not everyone can perform higher level medical procedures and/or administer life-saving medications, almost everyone can learn basic life support. Importantly, the first action in basic life support is to call for help. A similar mindset and plan may be applied to the active shooter situation. Different people have different attitudes and physical abilities; however, a generalized sense of situational awareness followed by at least a basic concept of action may provide a greater level of survival for all. It is also expected that some individuals may be capable of heroic acts while others will not—much of this may hinge upon prior education and training.

One specific training geared towards the active shooter incident is ALICE training (Alice Training Solutions, 2025). The ALICE training regimen is deemed the "civilian active shooter response training, delivered with a trauma-informed approach" (Alice Training Solutions, 2025). ALICE training briefly stands for: Alert, Lockdown, Inform, Counter, and Evacuate. Importantly, this training does not suggest fighting back in any way (Alice Training Solutions, 2025). However, it must be noted that in 2021, four of the 61 active shooters were killed by civilians involved in the active shooter situation (U.S. Department of Justice, 2021). Jasani et al. (2021) reviewed the critical nature of preparedness as proper training of staff. This mirrors that of medical response training. Due to the extreme and unpredictable nature of active shooter incidents, mock active shooter exercises are considered the most effective means of training by many (U.S. Department of Homeland Security, 2008). Jasani et al. (2021) noted that the most common type of active shooter training in their study was online, followed by lectures. While any training is better than none, preparedness for an active shooter incident necessitates specialized training. A first priority should always be to active emergency services. The scenario of the active shooter incident in the emergency department is unique in that there almost always exists at least one law enforcement officer in the vicinity. This is in typical contrast to most other soft target settings.

Our survey indicated that a majority of participants did not have any active shooter training at their place of employment. While this is not surprising, it is expected that preparedness is likely the single most useful resource in such a situation. Even if nothing more than mental resilience, some form of preparedness and training offers those caught on the site of an active shooter location a framework to work within. The ability to simply remain calm allows individuals to think more clearly, observe his/her surroundings more effectively, and overall be best able to mitigate a difficult situation.

Individuals would be prudent to make efforts to practice situational awareness, such that he/she might notice something or someone out of the ordinary and a potential risk. Simple practices such as this could result in life-or-death outcomes. Familiarity with workplace layouts, to include safe places and exit routes are also recommended to be aware of. Additionally, silent alarms can be easily installed and allow safe and immediate calls for help to authorities. While not advocated except for those who practice extensive training and familiarity, some individuals may elect to make a stand in such a situation. There are reports of successful outcomes whereby armed civilians are able to stop an active shooter; however, this would be considered a last resort as outcomes could be highly variable.

Conclusion

The active shooter scenario has become more familiar in time, with an increasing number of situations occurring in soft target locations. Familiarity with the active shooter concept and a strong sense of situational awareness are key factors in individual preparedness for such situations.

Joshua Lane holds an MD with board certification in Dermatology, Mohs micrographic surgery, and Disaster Medicine. He also has a PhD in Counterterrorism. His primary research in the defense sector pertains to counterterrorism medicine and active shooter mitigation. Additional research includes ballistic and optical targeting. He welcomes opportunities for continued research and collaboration.

References

Akiyama, C. (2020). "Surviving an active shooter incident in the intensive care unit." *Crit Care Nurse! 43*(1): 308.

ALICE Training Solutions. (2025). "What is ALICE Training?" Last modified 2025. https://www.alicetraining.com/.

Anklam, C., Adam, K., Sharevski, F., & Dietz, J.E. (2015). "Mitigating active shooter impact: Analysis for policy options based on argent/computer-based modeling." *J Emerg Manag 13*(3): 201-216.

Argintaru, N., Li, W., Hicks, C., White, K., McGowan, M., Gray, S., & Petrosoniak, A. (2020). "An Active Shooter in Your Hospital: A Novel Method to Develop a Response Policy Using In Situ Simulation and Video Framework Analysis." *Disaster Med Public Health Prep 9*(1): 1–9. doi: 10.1017/dmp.2019.161.

Clark, K.R. (2019). "Implementing an Active Shooter Policy and Training Program." *Radiol Technol 90*(4): 407–409.

Doherty, M. (2016). "From Protective intelligence to threat assessment: Strategies critical to preventing targeted violence and the active shooter." *J Bus Contin Emer Plan 10*(1): 9–17.

Federal Bureau of Investigation. (2025). "Active Shooter Safety Resources." Effective April 20, 2025. https://www.fbi.gov/how-we-can-help-you/safety-resources/active-shooter-safety-resources.

Federal Bureau of Investigation. (2023). "Active Shooter Incidents in the United States in 2023. Effective April 20, 2025. https://www.fbi.gov/file-repository/reports-and-publications/2023-active-shooter-report-062124.pdf/view.

Ferreira, C., Ribeiro, J., Almada, S., Rotariu, T., & Freire, F. (2016). "Reducing impacts from ammunitions: A comparative life-cycle assessment of four types of 9 mm ammunitions." *Sci Total Environ 1*(1): 566–567. DOI: 10.1016/j.scitotenv.2016.05.005.

Downs, S. (2015). "Active shooter in educational facility." *J Emerg Manag 13*(4): 303–326. DOI: 10.5055/jem.2015.0244.

Giwi, A., Milsten, A. Vieira, D. Ogedgbe, C. Kelly, K., & Schwab, IL. (2020). "Should I Stay or Should I Go? A bioethical analysis of healthcare professional and Healthcare Institutions' Moral Obligations During Active Shooter Incidents in Hospitals – A Narrative Review of the literature." *J Law Med Ethics 48*(2): 350–351.

Glasofer, A. and Laskowski-Jones, L. (2019). "Active shooter incidents: Awareness and action." *Nurs Manage 50*(3): 18–25.

Green, D. D. (2013). "Exploring Police Active Shooter Preparedness in Michigan: A Grounded Study of Police Preparedness to Active Shooter Incidents, Developing a Normative Model." *ScholarWorks at WMU.*

Greenberg, S. F. (2007). "Active shooters on college campuses: conflicting advice, roles of the individual and first responder, and the need to maintain perspective." *Disaster Med Public Health Prep 1*(1): S57-S61. DOI: 10.1097/DMP. 0b013e318149f492.

Goralnick, E. & Walls, R. (2015). "An active shooter in our hospital." *Lancet 385*(9979): 1728.

Guastello, S. J., Bednarczyk, C., Hagan, R., Johnson, C., Marscisek, L., McGuigan, L. & Peressini, A. F. (2022). "Team Situation Awareness, Cohesion, and Autonomic Synchrony." *Hum Factors 1*(1): 187208221118301. DOI: 10.1177/0018 7208221118301.

Inaba, K., Eastman, A. L., Jacobs, L. M., & Mattox, K. L. (2018). "Active-Shooter Response at a Health Care Facility." *N Engl J Med 379*(6): 583–586.

Jacobs, L. M., McSwain, N., Rotondo, M., Wade, D. S., Fabbri, W. P., Eastman, A., Butler, F.K., Sinclair, J., Joint Committee to Create a National Policy to Enhance Survivability from Mass Casualty Shooting Events. (2013). "Improving survival from active shooter events: the Hartford Consensus." *Bull Am Coll Surg 98*(6): 14–16.

Jasani, G., MacNeal, J., & Hirshon, J. M. (2021). "Emergency Department Active Shooter Training: A Survey of Current Practices in 2020." *Am J Dis Med 16*(4): 263–269.

Jones, J., Kue, R., Mitchell, P, Eblan, G., & Dyer, K. S. (2014). "Emergency medical services response to active shooter incidents: provider comfort level and attitudes before and after participation in a focused response training program." *Prehosp Disaster Med 29*(4): 350–357. DOI: 10.1017/S1049023X14000648.

Keep, J. J. (2018). "Active threat response: Building a resilient community." *J Bus Contin Emer Plan 12*(2): 119–132.

Kusulas, M. P., Drenis, A., Cooper, A., Fishbein, J., Crevi, D. Etess, M. S., & Bullaro, F. (2022). "Code Green Active" Curriculum: Implementation of an Educational Initiative to Increase Awareness of Active Shooter Protocols Among Emergency Department Staff. *Pediatr Emerg Care 38*(8): e1485-e1488. DOI: 10.1097/PEC.0000000000002666. Epub 2022 Mar 1.

Lesser, K. A., Looper-Coats, J., & Roszak, A. R. (2019). "Emergency Preparedness Plans and Perceptions Among a Sample of United States Childcare Providers." *Disaster Med Public Health Prep 13*(4): 704–708. DOI: 10.1017/dmp.2018.145.

Kelen, G. D., Catlett, C. L., Kubit, J. G., et al. (2012). "Hospital-based shootings in the United States: 2000 to 2011." *Ann Emerg Med 60*(6): 790–796. DOI: 10.1016/j.annemergmed.2012.08.012.

Martindale, M. H., Sandel, W. L., & Blair, J. P. (2017). "Active-shooter events in the workplace: Findings and policy implications." *J Bus Contin Emer Plan 11*(1): 6–20.

McKenzie, N., Wishner, C., Sexton, M., Saevig, D., Fink, B. & Rega, B. (2020). "Active Shooter: What Would Health Care Students Do While Caring for their Patients? Run? Hide? Or Fight?" *Disaster Med Public Health Prep 14*(2): 173–177.

Moore-Petinak, N., Waselewski, M., Patterson, B. A., & Chang, T. (2020). "Active Shooter Drills in the United States: A National Study of Youth Experiences and Perceptions." *J Adolesc Health 67*(4): 509–513.

Morris, L.W. (2014). "Three steps to safety: developing procedures for active shooters." *J Bus Contin Emer Plan 7*(3): 238–244.

Palestis, K. (2016). "Active Shooters: What Emergency Nurses Need to Know." *J Forensic Nurs 12*(2): 74–79. DOI: 10.1097/JFN.0000000000000113.

Price, S. S., Stricker, J. C., Ridenour, W. L., & Parker, R. D. (2020). "Proactive safety awareness and violence prevention training for professional students." *J Dent Educ 84*(6): 712–717. DOI: 10.1002/jdd.12137.

Sanchez, L., Young, V. B., & Baker, M. (2018). "Active Shooter Training in the

Emergency Department: A Safety Initiative." *J Emerg Nurs* 44(6): 598–604.

Sawyer, J. R. (2015). "How to avoid having to run–hide–fight." *J Healthc Prot Management 31*(2): 15–22.

Schonfeld, D. J. Melzer-Lange, Hasikawa, A. N., & Gorki, P.A. (2020). Council on Children and Disasters, Council on Injury, Violence, and Poison Prevention, Council on School Health 2020. *Pediatrics 146*(3): e2020015503. DOI: 10.1542/peds.2020-015503.

Scott-Herring, M. (2022). "Active Shooter Preparedness: Is Your OR Ready?" *AORN J 115*(6): 546-551. DOI: 10.1002/aorn.13691.

Thompson, A., Price, J. H., Mrdjenovich, A. J., & Kubchandani, J. (2009). "Reducing firearm-related violence on college campuses–police chiefs' perceptions and practices." *J Am Coll Health 58*(3): 247–254. DOI: 10.1080/07448480903295367.

U.S. Department of Homeland Security. (2008). "Active Shooter: How to Respond." Last modified October 2008. https://www.dhs.gov/xlibrary/assets/active_shooter_booklet.pdf.

U.S. Department of Homeland Security. (2019). "Security of Soft Targets and Crowded Places – Resource Guide." Last modified April 2019. https://activeshootersurvivaltraining.com/shared-files/9236/19_0424_cisa_soft-targets-and-crowded-places-resource-guide.pdf.

U.S. Department of Justice. (2018). "A Study of Active Shooter Incidents in the United States between 2000 and 2013." Effective September 16, 2013. https://www.amazon.com/Active-Shooter-Incidents-United-Between/dp/1986109879.

U.S. Department of Justice. (2021). "Active Shooter Incidents in the United States in 2021." Effective May 2022. https://www.fbi.gov/file-repository/active-shooter-incidents-in-the-us-2021-052422.pdf/view.

U.S. Secret Service. (2019). "Protecting America's Schools: A U.S. Secret Service Analysis of Targeted School Violence." Effective November 2019. https://www.secretservice.gov/sites/default/files/2020-04/Protecting_Americas_Schools.pdf.

Wallace, L. N. (2021). "Perceptions of Active Shooter Prevention and Preparation Strategies in Pennsylvania: Links to Self-Protective Behavior." *J Prim Prev 42*(1): 5–25.

Wallen, M. F., Drone, E., Lee, J., & Ganti, L. (2022). "Assessment of Emergency

Department Staff Awareness of Policy and Expert Opinion Protocol Regarding Active Shooter Events." *Disaster Med Public Health Prep 29*(1): 1–4. DOI: 10.1017/dmp.2022.116.

Wax, J. R., Cartin, A., Craig, W. Y., et al. (2019). "U.S. acute care hospital shootings, 2012–2016: A Content analysis study." *WOR* 2019; 64(1): 77–83.

Weisbrot, D. M. (2020). "The Need to See and Respond": The Role of the Child and Adolescent Psychiatrist in School Threat Assessment." *J Am Acad Child Adolesc Psychiatry 59*(1): 20–26. DOI: 10.1016/j.jaac.2019.09.001.

Book Review: *Terrorist Minds: The Psychology of Violent Extremism from Al-Qaeda to the Far Right*

Horgan, John. *Terrorist Minds: The Psychology of Violent Extremism from Al-Qaeda to the Far Right*. Edited by Bruce Hoffman. Columbia University Press (New York), 2024. ISBN: 978-0-23-119838-7. 248 pgs. $28.40

Reviewed by Dr. Don Meyerhoff
American Military University

John Horgan, renowned psychologist and researcher of terrorism topics, presents his latest publication, *Terrorist Minds*, which not only examines the motivations/ideologies linked to terrorist groups, but also examines the mindset, character, and influential environmental factors associated with terrorism. He further assesses the contemporary state and related challenges, addressing the study of this popular and timely field. The text is written for academics and laymen and is easily digested by a wide range of readers.

Horgan initially examines the topic of terrorism from a constructive analytical perspective to aid the reader in understanding fundamental elements that comprise the complex phenomena. Through an adroit use of cases, the author provides examples and context that point to the need for careful analysis of both the victim and the victimizer. Horgan also stresses the importance of examining and understanding the researcher's role in the analytical process, keeping tabs on any possible influencing biases. Additionally, it is noted that terrorism can be viewed as killing for a purpose, often embracing bigger worldviews or objectives outside of personal or criminal enterprise. This, coupled with state-based violence in the name of justified government responses and the protection of freedoms, only furthers the confusion and lack of certainty when attempting to understand or define terrorism in general.

The author discusses globalization, the resultant flow of information, and how they impact modern terrorism. These acts of violence committed in a far-flung country or region can, and do, influence those committed several oceans and continents away, thanks to the timely transfer of information over a multitude of modes and channels. Horgan provides numerous examples to provide the reader with clarity and context. He is careful not to limit the narrative to typical region-based examples of terrorism. Instead, he offers examples and discussions associated with diverse types of terrorism, such as that perpetrated by the more contemporary incel radicals or the "troubles" involving the Irish Republican Army.

Also addressed is the need for ideological motivation as a fundamental component in the terrorism equation. Through multiple contemporary

doi: 10.18278/si.10.1.10

examples and analyses, the author describes the difference between acts of violence and terrorism, a position that often confuses many laymen. The disparity in interpretation and how it impacts relevant statistical reporting is discussed, as well as the impact of biased media reporting methods. Inconsistencies in how the judicial system deals with different types of groups of convicted offenders further provide insight into how the perception of terrorism impacts society in general.

Horgan's latest work also examines various factors such as age, education, and gender (among others) and how they are represented in the composition of terrorism. Through the use of several examples and qualified studies, the author demonstrates the rather diverse and inconsistent manner of the modern terrorist construct. The text stresses the importance of understanding each group's customs as well as membership and role requirements, which can be uniquely specific and differentiate a group or type from each other. Equally interesting is the author's discussion of how diversity among different terrorist groups may vary widely across organizations.

Halfway through Horgan's latest work, motivation and the perpetrator's perception of wrong or right are discussed, referencing attackers with varying base ideologies. Horgan then discusses his own theory, IED, which addresses motivational factors associated with involvement, engagement, and disengagement, as well as those of other noted professionals such as Gavin

Bailey and Phil Edwards. Various approaches to justification for acts of violence are also presented with examples and sources.

Several popular models of radicalization and the interpretations of other noted researchers and professionals are presented. Studies and statistics throughout the section support differences in the types of radicalization models and the level and type of outward manifestations. The popular notion of multiple contributing factors in the shape of exposure and vulnerability needed to formulate a radicalized viewpoint is identified.

Mental illness or medical conditions and their contribution to a terrorist mindset are analyzed. Numerous factors and opinions regarding a causal relationship between the two are provided to the reader, along with numerous positions established and supported through formal studies. The effectiveness of analytical tools is also discussed. Other cognitive issues, such as the mindset of the terrorist and the use of tactics such as dehumanization, are presented with several positions as well as supporting statistics.

Recruitment and socialization methods are presented through examples and models, including those co-developed by Horgan in earlier research. Other noted researchers' approaches that use a categorical approach, such as that posited by Michael Hogg, are also discussed. Tactics used to develop the recruit and lead them to incremental progression and development towards radicalization are presented. These techniques and others, such as isolation from family and friends, adequately bond the recruit to the organization at a fundamental level.

Conversely, the author also addresses departure from terrorism and reintegration into society. Separation tactics and anxieties associated with deciding to part ways with a group or organization are presented with examples. A portion of the discussion is about the effect and impact of prison on a captured or convicted terrorist and how custody transforms the individual. Rationales for shifting cognitively from one role to another are discussed with several alternative hypotheses. Additionally, recidivism rates are presented and compared to other "criminal" offender types with some interesting findings.

Horgan provides the reader with insights gained from personal experiences obtained from communicating and interviewing with ex-terrorists both in and out of custody. Lessons learned and recommendations for various methods of approach and data col-

lection are discussed. Insights regarding biases and honesty on the interviewee's part are touched upon. Tools and suggestions for navigating the challenges associated with establishing contact and conducting interviews with those once involved with terrorism from several noted researchers provide the reader with ideas and concepts that may be useful in developing their future research projects.

The author wraps up this work by appealing to researchers and those controlling access to potential interviewees. Horgan identifies the importance of continuing research in numerous ways using the best available resources, thus enriching the available academic information pool.

Of note is the relatively large section of references and notes at the end of the text. Dr. Horgan provides the reader with a rich array of quality sources for many topics covered in his book. These can be excellent jumping-off points for further depth in specific subjects or topic areas for further research.

Related Titles from Westphalia Press

The Zelensky Method
by Grant Farred

Locating Russian's war within a global context, The Zelensky Method is unsparing in its critique of those nations, who have refused to condemn Russia's invasion and are doing everything they can to prevent economic sanctions from being imposed on the Kremlin.

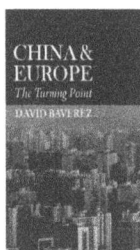

China & Europe: The Turning Point
by David Baverez

In creating five fictitious conversations between Xi Jinping and five European experts, David Baverez, who lives and works in Hong Kong, offers up a totally new vision of the relationship between China and Europe.

Masonic Myths and Legends
by Pierre Mollier

Freemasonry is one of the few organizations whose teaching method is still based on symbols. It presents these symbols by inserting them into legends that are told to its members in initiation ceremonies. But its history itself has also given rise to a whole mythology.

Resistance: Reflections on Survival, Hope and Love
Poetry by William Morris, Photography by Jackie Malden

Resistance is a book of poems with photographs or a book of photographs with poems depending on your perspective. The book is comprised of three sections titled respectively: On Survival, On Hope, and On Love.

Contests of Initiative: Countering China's Gray Zone Strategy in the East and South China Seas
by Dr. Raymond Kuo

China is engaged in a widespread assertion of sovereignty in the South and East China Seas. It employs a "gray zone" strategy: using coercive but sub-conventional military power to drive off challengers and prevent escalation, while simultaneously seizing territory and asserting maritime control.

Frontline Diplomacy: A Memoir of a Foreign Service Officer in the Middle East
by William A. Rugh

In short vignettes, this book describes how American diplomats working in the Middle East dealt with a variety of challenges over the last decades of the 20th century. Each of the vignettes concludes with an insight about diplomatic practice derived from the experience.

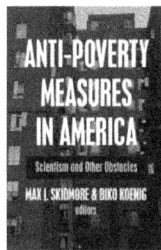

Anti-Poverty Measures in America: Scientism and Other Obstacles
Editors, Max J. Skidmore and Biko Koenig

Anti-Poverty Measures in America brings together a remarkable collection of essays dealing with the inhibiting effects of scientism, an over-dependence on scientific methodology that is prevalent in the social sciences, and other obstacles to anti-poverty legislation.

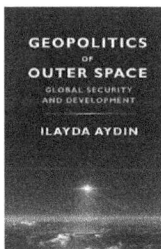

Geopolitics of Outer Space: Global Security and Development
by Ilayda Aydin

A desire for increased security and rapid development is driving nation-states to engage in an intensifying competition for the unique assets of space. This book analyses the Chinese-American space discourse from the lenses of international relations theory, history and political psychology to explore these questions.

westphaliapress.org

American
Public
APU University

American
Military
AMU University

www.ingramcontent.com/pod-product-compliance
Lightning Source LLC
Chambersburg PA
CBHW081407270326
41931CB00016B/3407